BIOCHEMISTRY UNITS FOR THE HIGH SCHOOL BIOLOGY TEACHER

Other books by the author:
  Handbook of Modern Experiments for High School Biology

Biochemistry Units for the High School Biology Teacher

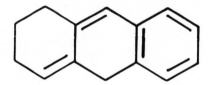

Adelaide Hechtlinger

PARKER PUBLISHING COMPANY, INC.
WEST NYACK, NEW YORK

© 1973, *by*

PARKER PUBLISHING COMPANY, INC.

West Nyack, New York

Library of Congress Cataloging in Publication Data

Hechtlinger, Adelaide.
  Biochemistry units for the high school biology
teacher.

    1. Biological chemistry--Study and teaching
(Secondary) I. Title.
QP518.H4   574.1'92'0712   73-175
ISBN 0-13-076471-X

Printed in the United States of America

How This Book Will Help
the High School Biology Teacher

B iology today is not the same biology that was taught fifty, twenty-five, ten, or even five years ago. With new instruments, techniques, and other research methods, biology changes its character rapidly. Just as the teacher accepts a theory into the body of materials, a new theory is uncovered and a new set of facts must be relearned and retaught. We would find many of these new discoveries less difficult to learn or teach if there was a proper foundation of chemistry.

In this book the author covers the subject of biochemistry on a level elementary enough to offer *practical* value to teachers. The result will enable readers to feel more secure in their knowledge of the relationship of chemistry to biology.

At the present time one cannot fully understand the workings of living organisms without some basic knowledge of chemistry. It is necessary for teacher and student alike to have information in this field. Yet the field of biochemistry is so new that often the teacher is not fully acquainted with materials in the field. It is only in the past few years that many schools require the biology student to take chemistry courses.

Since the field of biochemistry is expanding so rapidly, there is need for a book that emphasizes the practical aspects. Biochemistry books, on the whole, are large and unwieldy, and for most teachers, the large books are frequently unsatisfactory and unnecessary.

This book emphasizes material that can be applied to the classroom situation. The various classes of food are discussed as well as the mechanisms of digestion that prepare these foods for use by the organisms. Fat-soluble vitamins and hormones are also included in the discussion.

No course is complete without laboratory exercises. The author is a firm believer in an old adage, which can be paraphrased for the science student by saying, "One laboratory exercise is worth a thousand words." You will have more affinity for the material if you use laboratory exercises and demonstrate some of the chemical reactions. You will be able to visualize some of the reactions that actually take place within a living cell or organism.

Each chapter deals with a particular phase in the study of biochemistry, such as proteins, carbohydrates, lipids, and enzymes. There is a short review at the beginning of each unit to emphasize those facts that are important, including a number of short demonstrations to be shown to the students as well as a group of laboratory exercises to be found at the end of the chapter. These exercises are not open-ended, but of the cookbook type. Directions are concise and state what should be expected. The field of biochemistry is still new enough so that many people do not know what to look for.

In addition to a variety of laboratory exercises, a small number of visual aids are helpful, such as overhead transparencies, filmstrips, single-concept films, and 16mm films. A list of these aids is included at the end of each unit, and the Appendix indicates the sources of supply for these visual aids.

—Adelaide Hechtlinger

Contents

UNIT 17   MOLECULAR GENETICS *(cont.)*

Biochemistry and the Cell

UNIT 1

Not too long ago, the study of the cell was quite simple; you were taught that the cell contained a nucleus, cytoplasm, a cell membrane, and if it was a plant cell it contained a cell wall and several vacuoles. It is no longer that simple. The electron microscope has shown that there are many other parts to a cell, and upon further study it was found that the processes carried on within the confines of the cell were not all that simple.

Today, in order to fully appreciate the workings of the cell, whether it be a one-celled organism or part of a multicellular organism, you need a working knowledge of biochemistry. The human body contains approximately one hundred million million cells. The chemical processes carried on within a cell are practically unbelievable, as you will see when the subject of biochemistry progresses.

Before anything more is said, it would probably be best that you review the parts and workings of a single cell. Then you can tackle the intricate processes.

(If you have time, review the cell by doing a laboratory exercise on the plant cell [onion] and the animal cell [cheek] as seen under a regular microscope. You also could set up several microscopes with slides of the onion and cheek cells to be used for demonstrations, instead of devoting an entire period or more to a laboratory exercise.)

When biologists looked at a cell under the electron microscope, they found many structures that were never seen previously. Under an ordinary microscope the membrane surrounding
the cell seemed to be but one layer thick. The electron microscope
showed that the membrane consists of three separate layers; the
inner and outer ones were composed of proteins, and the middle
layer composed of lipids. The proteins in the membrane allow the
membranes to expand or contract according to the chemistry of
the environment around the cell, while the lipid layer is concerned
mainly with allowing the passing of fat-solvent materials in and
out of the cell. Plant cells have massive walls of cellulose
reinforcing the cytoplasmic membrane.

As for the cytoplasm, the electron microscope has shown the
presence of numerous structural bodies, which have been isolated
from the nucleus. These are the mitochondria, fat globules,
endoplasmic reticulum, the Golgi complex, the centrosome, and
the nucleus. It was once thought that the cytoplasm was a
homogeneous syrupy solution, but today it is believed to be an
intricate network of elongated protein molecules responsible for
many of the cell's functions.

In the plant there are chloroplasts as well as the other bodies.
However, you need not go into detail because chloroplasts are
most important when you go into the entire topic of photosynthesis.

(For demonstration purposes, use the models of the plant
and animal cells as seen under the electron microscope or use
transparencies or charts, since it is of utmost importance that the
student understand you as you discuss the various parts of the
cell.)

The mitochondria are called the "power plants" or dynamos
of the cell since they are the "reactors" in which sugar products
are burned and their energy used to form adenosine triphosphate
(ATP), the chemical that is the immediate source of energy for
most vital processes. The mitochondria are fluid-filled vessels
enveloped in a thin double membrane, with the inner layer folded
to form partitionlike ridges, crests, or shelves. The wall consists of
a double membrane made up of single layers of lipid and protein
molecules. Within the protein layers are found special respiratory
enzymes, vitamins, and minerals needed for the oxidative metabolism of foodstuffs. There are estimated to be about twenty-five
different enzyme systems in the mitochondria, and that as many

as 2,000 duplicates of each enzyme system can exist simultaneously.

The mitochondria are also concerned with the breaking down of fat to yield energy and the building up of fat when more sugar is available than needed for immediate use. Fat is the principal means by which the body stores reserve supplies of energy. It is probably this association of mitochondria with fat metabolism that accounts for the finding of fat droplets so close to the mitochondria.

When you study the models of the cell closely, notice that throughout the cytoplasm there is a mass of very fine tubular filaments. These are called endoplasmic reticula, and clinging to the outer surface of each endoplasmic reticulum are dots of ribonucleoprotein (RNP), called ribosomes. Roughly spherical in shape, they contain protein and a relatively high proportion of one form of nucleic acid, ribonucleic acid (RNA). The ribosomes are commonly believed to be the primary sites of protein synthesis within the cell.

Another inclusion in the cytoplasm is the Golgi apparatus or complex. This structure is well developed in nerve, germ, and secretory cells. For a long time there has been a great deal of controversy concerning this structure. These minute tubules appear to be filled with fluid and resemble the endoplasmic reticulum. Thus it is felt that this might be a storage area which can quickly pass its material to various parts of the cell as the demands arise.

The lysosomes are minute particles of dense and granular material, bound by protein membranes and containing many enzymes which are hydrolytic in function. The digestive enzymes (proteins) break down large nutrient molecules into smaller ones to be oxidized by the mitochondrial enzyme system. If the membrane of the lysosome is ruptured, the release of its contents causes rapid lysis of the cell.

The centrosome, or cell center, is an area of dense protoplasm which becomes plainly visible as the cell is about to divide. Within the centrosome are a pair of rod-shaped granules, the centrioles. The centrosomes are replicated during mitosis and appear to be primarily concerned with this process. Since they are made up of fibers and resemble cilia and bacterial flagella, it might be that the centrosomes are composed of contractile proteins.

The nucleus, usually located in or near the center of the cell,

contains the chromosomes, responsible for transmitting the hereditary pattern from one generation to the next. The chromosomes
contain the vitally important filaments of chromatin, to which one
form of nucleic acid (deoxyribonucleic acid or DNA) of the cell is
confined with its characteristic proteins. It is believed that the
nuclear membranes have holes through which the materials pass. It
may be that the nucleus exerts its regulatory effects on cell
processes when the ribonucleoprotein migrates from the nucleus
to the cytoplasm.

Within the nucleus can be seen the nucleolus. This body
disappears during cell division and reappears at the end of the
division. The nucleolus is rich in proteins and RNA, and may be
the site of the protein and RNA synthesis.

Now that you have considered the various structures of the
cell with a brief description of their composition and functions,
you will be ready to go into biochemical detail of the cell and its
various life processes.

A cell lives by virtue of its composition and organization.
Both are unique: the composition in large part because of the
universal presence of certain classes of very large molecules,
so-called macromolecules, the largest and most complex in all
chemistry; and those are responsible for many of the most
distinctive features of cellular organization and behavior.

All macromolecules fall into three great classes—polysaccharides, nucleic acids, and proteins—common to all cells and
sharing common properties within each class.

Each type of macromolecule is composed of a limited
number of repeating subunits, bound together to form long chains.
Sugars form polysaccharides, nucleotides form nucleic acids, and
amino acids form proteins.

In all types of macromolecules, the subunits are bound to
one another through the same device, the elimination of a
molecule of water between each pair. Conversely, every macromolecule may be broken down into its subunits by the reverse
process, the insertion of a molecule of water between each pair.
The latter process is called hydrolysis.

(If you have time, there are several good films that can be
shown to the students to enhance their complete understanding of
the cell as we know it. Another good class aid is the microviewer,
with the strip of the ultrastructure of animal cells and another
strip of cell structure. There are eight slides on the strip, with the
explanations and questions for each slide.)

At this point, before actually beginning the biochemistry, I would suggest that you and your class do the basic laboratory exercises on paper and thin-layer chromatography. Chromatography is used many times in later laboratory exercises and a basic understanding of the concepts behind this technique could be of value. There is also a simple exercise on electrophoresis, another recent technique used in laboratories.)

The laboratory exercises on the plant and animal cells, the onion and cheek cells respectively, will not be given here since the exercises are so basic that every biology teacher should know them.

Laboratory Exercise: Paper Chromatography

Aim: How to make a very simple chromatogram.

Chromatography is a rapid method of separating and identifying the components of a mixture without destroying it. It has become an important tool for the research scientist and is the basis of exacting techniques used in such recent discoveries as the uncovering of parts of the genetic code and the determination of its structure.

The first type of chromatography to be used was paper chromatography, a crude form of which will be done in this exercise. The mixture to be studied is first permitted to be absorbed as a spot on the filter paper. A selected solvent then passes through the spot and ascends the paper by capillary action. As the solvent passes through the absorbed mixture, some substances dissolve more rapidly than others, and thus travel along with the solvent at a faster rate. They thus travel farther up the paper before they are absorbed by the paper again at a higher level. When the paper is removed from the solvent, the various components of the mixture may be present as spots at different heights. If you already know how high the various substances ascend in this solvent system, you may identify the unknown substances in the mixture. Descending and circular methods of paper chromatography are also employed.

The heights of each of the spots can be measured from the point of application to the level to which it has moved within a specified time. This value is called the R_f value.

$$Rf = \frac{\text{height of the spot}}{\text{height of solvent front}}$$

There are several other types of chromatography in use today, such as Thin-Layer Chromatography (TLC), which you will be hearing about later on.

Materials Needed: A piece of filter paper one inch wide and about eight inches long, a glass rod, stapler, beaker, mixture of food colors.

Procedure: (1) Put one end of the strip of filter paper around a glass rod and staple it in place. (2) Make a mixture of food colors by placing several drops of each of the food colors into a beaker and then mix thoroughly. (3) Add a little water to the mixture to dilute slightly. (4) Lay the glass rod across the top of the mixture in such a manner that the strip of paper is just below the surface of the mixture. (5) Watch what happens. Within a few minutes, the solution will rise on the paper and gradually each color that you put into the mixture will separate and form a separate band across the width of the strip. Every color that you place in the mixture will be seen on the filter paper as a separate band and in a different location.

Thus you see that with this very crude method of chromatography you will be able to separate the components of a mixture. The more refined methods will be used to separate other mixtures as you go along.

Laboratory Exercise: Thin-Layer Chromatography

Aim: To become familiar with the technique of thin-layer chromatography.

As was previously mentioned, chromatography is a rapid method of separating and identifying the components of a mixture without destroying them. In paper chromatography, the mixture to be studied is first permitted to absorb as a spot on the filter paper.

We shall use a new chromatographic procedure known as thin-layer chromatography (TLC). Instead of paper, the absorbing surface is a layer of Silica Gel G (Merck) on a microscope slide.

For the research worker, TLC is superior to paper chromatography in certain kinds of analysis. For us the advantage is speed.

There are prepared chromatogram sheets for TLC but in most cases the material has to be activated before using, whereas this method of making your slides does not require such preparation.

We shall only give you in this exercise the method of preparing the microchromatic plates. The various materials that can be used with these slides will be forthcoming at a later time.

Materials Needed: Slides, Coplin jar, Silica Gel G (Merck), chloroform, methanol.

Procedure: (1) Place 21 grams of Silica Gel G powder into a container, such as a Coplin jar, and then add a mixture of 33.5ml chloroform and 16.5ml methanol. Stir vigorously for several minutes to form a slurry. When not in use, the jar should be covered with a greased top to prevent evaporation. The slurry may become thick because of evaporation and additional solvent may be added to keep the slurry at the proper consistency. (2) To dip the slides, new slides are not necessary as long as the slides have been washed and cleaned in the usual manner in any detergent and then air-dried. Place two slides together back-to-back so that one extends slightly over the other at the top, and insert both in the slurry to a depth of about 1 cm from the top of the slides. Hold the slides at the top while doing this, withdraw the pair of slides, and scrape the bottom edge of the slides on the rim of the Coplin jar to remove excess silica gel. Allow to air-dry for a few seconds. (3) Holding the slides firmly, pull the upper slide away from the lower slide and place them, silica side up, on a paper towel. Since these slides were prepared with a nonaqueous solvent, they are ready for use at once without further drying in an oven. After the slurry has been mixed, enough microchromatic plates can be prepared for an entire class in a few minutes. (4) Coated slides can be used after two weeks if left undisturbed. Jarring them will cause the dried gel to disintegrate.

Note to teacher: There will be enough slides prepared for use by a class of twenty students.

Laboratory Exercise: Paper Electrophoresis

Aim: How to do a simple electrophoresis.

Electrophoresis is based on the principle that many substances in solution are dissociated into ions which have a characteristic positive or negative charge. These ions will migrate at a characteristic rate under the influence of an electric field. Actually electrophoresis is an incompleted form of electrolysis, since the desired products are not liberated at the electrodes but their movements are stopped simultaneously at some intermediate point between the electrodes.

Substances which ionize in solution have an electric charge, and when exposed to the two poles of an electric field each ion will migrate to one pole or the other, depending on the charge it carries. An ion with a negative charge will migrate to the positively charged anode, while the ion with the positive charge will move toward the negatively charged cathode.

The mechanics are very simple. The strip of filter paper is moistened with a suitable buffer solution and the excess buffer is removed by means of a blotter. The buffer solution is necessary to maintain reasonably constant conditions of pH during the experiment; otherwise, the resistance of the paper would be so great that no current would flow. A small drop of the test solution is placed on the center of the strip or a streak of the solution is applied along a pencil line on the filter paper. The paper strip is then placed between two buffer-containing electrode compartments, one end dipping into the anode compartment and the other into the cathode compartment. The electrodes are connected to a source of low-voltage current. The ionized substances of the mixture being tested are attracted to the electrode compartment of the opposite charge and start to move along the paper in that direction. The electrophoretic analysis is run until the components of the mixture have separated sufficiently. The current is then switched off, and the strip of paper removed and dried as quickly as possible in a hot oven or in front of an infrared lamp. The spots on the paper can be quantitatively evaluated in the manner described for paper chromatography.

Materials Needed: An electrophoresis apparatus and power pack to run it with all clips, etc. that are needed; a mixture of inks to use as a sample, containing 1 ml brown ink, 1 ml blue ink, 1 ml of green ink, 1 ml red ink; oxoid barbitone acetate buffer (which is made by weighing 6.6 grams of Oxoid Acetate buffer powder and adding it to 1 liter of distilled water, and mixing to dissolve, pH

8.6. Store in a stopper bottle until needed, but it is advisable to add 1 ml 5% thymol in isopropanol to a liter of buffer to act as a preservative.

Procedure: (1) Fill the electrophoresis tank as per instructions given to you with apparatus. Make certain that buffer is level in compartments and wipe any moisture from the top of the partitions. (2) Use a soft lead pencil to make identifying marks in a corner of the strip. (Mark a line for sample application across the strip in a position approximately 1/3 the length of the strip.) (3) Float the strip on the surface of the buffer in the tank. When the strip is wet, submerge it to complete wetting. (4) Remove the strip from the buffer and blot between clean sheets of filter paper to remove excess moisture. Handle the strip as little as possible and then only at the end. (5) Place the strip across the bridge gap, allowing approximately a 2 cm overlap of the strip at either side of the shoulder pieces. Position the strip in the tank. (6) Read instructions as to how to make the strip taut. (7) Plug in the tank and switch to "on" position, and equilibrate the buffer and strips for four minutes. (8) Turn switch to "off" position and remove cover to apply sample to strip. (9) Touch a clean capillary pipette to the ink mixture and allow the mixture to rise a short distance in the pipette. (10) Apply sample to strip in the form of a uniform streak along the previously marked line. Move the pipette back and forth across the strip to make the line equal. The line should be approximately 1/3 the distance of the bridge gap from cathode side. Allow approximately 0.5 cm margin on either side of the line of sample application. (11) Place cover on the tank and turn switch to "on" position. Allow to run for fifteen minutes. (12) Remove strips from the tanks as outlined in the general instructions with kit. (13) Blot the strip between clean sheets of filter paper to dry. (14) There will be a separation of ink pigments which can be observed in the course of the electrophoresis run.

AUDIO-VISUAL MATERIALS

The letters in parentheses following the audio-visual material are the code letters for the source of supply. The entire list will be found in the Appendix.

FILMS

>Biochemical Origin of Terrestrial Life (B)
>Biology in Today's World (K)
>Cell Biology: Life Functions (K)
>Cell Biology: Structure and Composition (K)
>Cells of Plants and Animals (K)
>From Primitive Gases to Amino Acids (Q)
>The Origin of Life—Chemical Evolution (J)

FILMSTRIPS

>The Cell (set of two): What Is a Cell?; How Cells Work (R)
>New Concepts of the Cell (C)

SINGLE-CONCEPT FILM LOOPS

>Cell Model: Animal (P)
>Paper Chromatography (D)
>Paper Electrophoresis (D)
>Thin-Layer Chromatography (D)

TRANSPARENCIES

>Animal Cell: Generalized (A), (F)
>The Cell Set—Transparencies of all ultraparts of the cell (F)
>Organelles (F)
>Plant Cell: Generalized (A), (F)

CHARTS

>Cell Structure (D), (G)
>Generalized Cell (F)

MODELS

>Animal Cell Plaque (A)
>Animal Cell Section (D)
>Cell Model, Animal (A), (D), (F), (G), (S)
>Cell Model, Plant (A), (D), (F), (G), (S)
>Centriole (D)
>Golgi Complex, Animal (D)
>Golgi Complex, Plant (D)
>Mitochondrion (A), (F), (G)
>Mitochondrion, Animal (D)
>Mitochondrion, Plant (D)
>Plant Cell Plaque (A)
>Plant Cell Section (D)

MICROVIEWERS

Cell Structure (M)
Ultrastructure of Animal Cells (M)

KIT

Electrophoresis Kit and Power Supply (D), (N)

Chemical Structure of Matter

UNIT 2

At this point, it would be best to discuss some of the basic concepts of matter and define a few of the more basic chemical terms.

As you know, matter is first in importance since it is anything which occupies space and has weight. Whatever you touch is matter. The air you breathe is matter. Anything which you can detect through your senses or by instruments is matter, whether it be living or nonliving.

Substance is a part of matter which differs from all other parts because of certain qualities. These qualities are known as properties. Substance is matter which is uniform throughout, such as water, while a material such as milk is not a substance because it is composed of several kinds of matter.

Matter may exist in any one of three states—solid, liquid, or gas. Remember that it is possible for most substances to readily change from one of these states to either of the others by means of heat, whether added or subtracted. You can readily show that water, a liquid, with the aid of heat becomes steam, a gas, and becomes ice, a solid, when heat is removed. Actually the fundamental composition of the water is not altered, only its physical state. The changes are completely reversible. This quality distinguishes water from other substances. It is a physical property, and the reversible change it refers to is a physical change.

Now, on the other hand, if you were to do an electrolysis of water, you would find that the water has separated into its components, hydrogen and oxygen; that would be a chemical change. The hydrogen and oxygen would not have the same appearance as the water did; the hydrogen and oxygen would have different properties. The change would be a chemical change and it would be the chemical property of water.

You can show that the dissolution of sugar or salt in water is a physical change because the removal of water by evaporation will yield the sugar or salt unchanged.

(You have several demonstrations and laboratory exercises that can be used at this point to emphasize these concepts. There are a number of audio-visual aids listed that would fit in with the topic at this point. Perhaps it might be a good idea to set aside one day to the showing of films, film loops, or filmstrips, while the transparencies and charts are used on a daily basis as part of the lesson. There are also several good kits that might be useful as laboratory exercises.)

ATOMIC STRUCTURE

Unless you have a fundamental knowledge of the particles that make up matter, you, as a biologist, might find some difficulty in understanding the role of matter in the operation of the total organism.

All matter, living and nonliving, consists of the same fundamental units, although they may be put together in various ways. Matter, be it gas, liquid, or solid, consists of one thing—atoms. Point out that the kinds of atoms found are few compared to the enormous number of substances.

Although at one time it was believed that the atom was indivisible, it is now known that the atom consists of many kinds of particles. You, however, will be concerned with but three.

All atoms have the same basic structural plan. The typical atom has a nucleus composed of two kinds of subatomic particles, protons and neutrons. Around this nucleus a third type of particle, the electron, travels at a very high velocity. The orbital path of an electron, or electrons, is called a shell. Each electron has its own orbit, which shifts about. Even when a number of such particles are in orbit around the nucleus, the shell formed is practically 100 percent space. The shell of an atom represents the average distance of a particular electron orbit, or several different orbits, from the nucleus. The electron is the only matter in the shell. The nucleus contains just about all the matter that the atom possesses. Between the nucleus and shell there is nothing but space.

It would be valuable to remember that the electron has a negative electrical charge, and that for every electrical charge outside the nucleus there is a positive charge inside the nucleus carried by the proton. When the number of protons inside the nucleus of an atom equals the number of its electrons, you would be correct in saying it is electrically neutral.

In simple atoms, protons account for about half of the total weight of the atom, while the neutrons account for the remaining portion of the atom's weight. The neutron, however, has no electrical charge and therefore is neutral; yet it still is an important particle in the nucleus of an atom.

The hydrogen atom is the only atom that has no neutron in its nucleus, just one lone proton. The hydrogen atom has one shell, since it has a single electron orbiting around the nucleus. The more complex atoms may have several shells, for they have many electrons arranged at different distances from the nucleus.

(There are transparencies which will explain this concept quite simply. Use them to help bring across the concept to the class.)

In biochemistry you should keep in mind that the electron is much more important than the proton or neutron. Biochemical processes in living things are taking place constantly. The chemical activity of an atom depends entirely on the number and arrangement of the electrons around the nucleus. Thus you can see why the chemical activity of the various fluids, tissues, and cells of an organism is of necessity a matter of electron behavior.

Since the electrons are so significant in chemical activity, here are a few simple rules of atomic structure that you should know. The first shell, closest to the nucleus, can contain no more than two electrons. The second shell is farther from the nucleus and can contain no more than eight additional electrons, although it may contain less than eight. If any atom has more than ten electrons it must have more than two shells.

Atoms also have *atomic mass* and *atomic number.* Many people are unable to distinguish between the two. The atomic mass is really a relative value. The atomic number is the position of a given kind of atom in a list of the 105 known kinds of atoms arranged, with a few exceptions, in order of increasing mass. The atomic number is always a whole number. The atomic mass of an atom is determined by using the number 12, the atomic mass of carbon, as a standard. Atomic mass can be found in the tables.

If you know the atomic mass and the atomic number of an atom, you can immediately know the structure of the nucleus of that atom. The atomic mass is numerically equal to the sum of the protons and neutrons. The atomic number of an atom is numerically equal to the number of protons it contains. Thus the difference between the atomic mass and the atomic number equals the number of neutrons present. With few exceptions, atoms generally form combinations with other atoms. This combination of atoms is called a molecule.

ELEMENTS

Frequently people cannot differentiate between atoms and elements. It might be of some help if you think of an atom as being a particle with a certain size and shape, while an element is a quantity of matter made up of atoms having the same atomic number. There are as many kinds of elements as there are kinds of atoms. For example, iron as an element is made up of particles called iron atoms, while sulfur as an element is made up only of particles called sulfur atoms. The elements which these two kinds of atoms comprise do not resemble each other.

The elements consist of molecules, which are the smallest part of the element to retain the properties of the element. Remember, also, that each element is unique in atomic structure and chemical properties as well as in its physical properties. Some elements are gaseous, such as hydrogen, helium, and oxygen; others are liquid, such as bromine and mercury. Most are solids, but each solid looks and acts differently from the others. This results from the fact that each atom is unique in its structure.

Some elements have several forms of the same kind of atom. The form may differ in mass, but the structure of the atoms is identical except for the number of neutrons. Chemists call these different atomic forms *isotopes*. Their atomic number is the same, so they are atoms of the same element. As you probably have read, radioactive isotopes are used in medicine and biological research. The isotopes are radioactive because their atoms break down spontaneously with a simultaneous release of great amounts of energy. This release of energy may have an effect on life.

(Since you are concerned with the role of elements in biology, the remainder of this discourse on elements is very important to you.)

Of all the 92 elements that occur in nature, only six are of major importance to the makeup of the organism. These six are found to be more than one percent of the total body weight for each organism. The elements are:

ELEMENT	% of Total Body Weight	Biological Importance
Hydrogen	10	The first four elements are the fundamental constituents of carbohydrates, fats, and proteins.
Oxygen	65	
Carbon	18	
Nitrogen	3	
Calcium	2	99% of body calcium is found as $Ca_3(PO_4)_2$, the major structural component of bone and teeth; also essential for blood coagulation and rhythmic activity of the heart.
Phosphorus	1.1	Indispensable for biochemical synthesis and energy transfer

There are thirteen other elements found in the human body, as follows:

ELEMENT	% of Total Body Weight	
Potassium	0.35	Principal intracellular cation essential for transport of nerve impulses, muscle contraction.
Sulfur	0.25	Constituent of some proteins and many other important biological compounds.
Sodium	0.15	Principal extracellular cation; proper balance and distribution of water in the body.
Chlorine	0.15	Principal intracellular and extracellular anion.
Magnesium	0.05	Required for the activity of many enzymes.
Iron	0.004	Most important ion; found in hemoglobin; concerned with oxygen transport and utilization.

Trace Elements

Zinc	0.0002	Required for certain enzyme activity.
Manganese	0.00013	Required for the activity of a number of enzymes; functions in animal reproduction.
Copper	0.0001	Essential constituent of vital oxidative enzymes; used in synthesis of hemoglobin.
Fluorine	0.0001	Minor constituent of some body structures, such as teeth.
Iodine	0.0001	Essential constituent of the thyroid hormone.
Molybdenum	0.0001	Required for certain enzyme activity.
Cobalt	0.0001	Structural component of vitamin B_{12}; deficiency results in anemia.

COMPOUNDS AND MIXTURES

If you look at a piece of granite with a hand lens, you can see that the granite is made up of several different crystalline forms; when you look at more than one piece of granite you will notice that the crystalline forms are in all the pieces of granite, but the amounts differ from piece to piece. Granite is not a compound but a mixture, since a mixture may have variable components and the components of a mixture may be separated by physical means, such as iron and sulfur. A mixture does not have to contain a constant composition of ingredients.

(At this point, do the mixture demonstrations and laboratory exercises. There also are several good films on the topic. Perhaps you should set aside a day to use the films.)

Compounds are formed when two or more dissimilar atoms combine. A molecule consists of two or more atoms combined into one particle. It is the smallest unit of a compound. The molecules of compounds retain the properties of the compound and consist of dissimilar atoms. The huge variety of substances in the world is due to the large number of chemical combinations possible. The property of a compound will differ from that of its individual components.

(The compound demonstrations should be done if they have not already been shown.)

Now you can understand why the formulas of compounds often include a subscript after some of the chemical symbols: water, H_2O; carbon dioxide, CO_2; glucose sugar, $C_6H_{12}O_6$. Each number tells how many atoms of the element help make up one molecule of the compound. Water molecules consist of two hydrogen atoms combined with one atom of oxygen. (The number 1 is understood without being written.)

Many common gases are written thus: oxygen as O_2 and hydrogen as H_2. Molecules of these gases always consist of two identical atoms.

There are two types of compounds, inorganic and organic. It should be explained to the students that *inorganic compounds* are those not directly associated with life. They seldom contain carbon and are of relatively simple composition. The four major classes of inorganic compounds are:

> *Oxides:* Compounds formed through combination of oxygen and other elements.
>
> *Acids:* Compounds consisting of a combination of hydrogen with one or more other elements. All acids produce hydrogen (H^+) ions when in solution.
>
> *Bases:* Combinations of hydrogen and oxygen with one other element or with a radical—produce hydroxide ions (OH^-) when in solution.
>
> *Salt:* Combination of the positive ion of a base and a negative ion of an acid.

The second type of compound is the *organic compound* which is associated with life, contains carbon atoms, and is usually of a very complex composition.

ORGANIC COMPOUNDS

The fundamental biological unit is known as the cell. The structure of the cell is based on three kinds of very large polymer-type molecules, known as carbohydrates, nucleic acids, and proteins. In addition to these kinds of structural units, animal and plant cells also contain fats and lipids, water, and a small amount of inorganic materials.

Basically, each of these organic compounds is found as one of several types:

Hydrocarbon: Contains only carbon and hydrogen; methane and ethane are examples and the key grouping is:

$$\begin{array}{c} H \\ | \\ H-C-H \\ | \\ H \end{array}$$

Alcohol: Simplest organic compound having oxygen as well as carbon and hydrogen in the structure of its molecule; methyl and ethyl alcohol are examples and the key grouping is:

$$\begin{array}{c} | \\ -C-O-H \\ | \end{array}$$

Aldehydes and Ketones: More complex organic compounds comprised of carbon, hydrogen, and oxygen. The presence of a double bond distinguishes them from that of the alcohols. The molecule of the compound must contain an oxygen atom that shares two pairs of electrons with a carbon atom. A double bond forms at that place in the molecule.

$$-C\overset{O}{\underset{H}{\diagdown}}$$

In the *aldehyde,* the key group is located at the end of the carbon chain, as formaldehyde.

In a ketone, the key grouping is not at the end of the carbon chain but usually an interior carbon, as in dimethyl ketone.

$$\begin{array}{c} | \quad\ | \\ -C-C-C- \\ |\ \ ||\ \ | \\ O \end{array}$$

Organic Acids: Organic compound that behaves like an acid. Acetic acid is an example and the key grouping contains two oxygen atoms.

$$C-O-OH$$

Amines: Organic compound that contains nitrogen in combination with hydrogen and carbon; methyl and ethyl amines are examples and the key grouping is:

$$-\overset{\displaystyle|}{\underset{\displaystyle|}{C}}-H\overset{\nearrow H}{\searrow_N}$$

All protein molecules are made of amino acids, and without protein molecules there is no organized life.

Ring Structures: Carbon atoms united in a ring rather than a chain; example is the benzene series—in living materials, purines and pyrimidines are in ring structures.

$$-\overset{\displaystyle C-C}{\underset{\displaystyle C-C}{C}}\overset{\diagdown\diagup}{\diagup\diagdown}C-$$

WATER

The most important and most abundant compound in living matter is water.

In man, the tissue with the lowest concentration is the dentine of the teeth (10 percent); the greatest concentration is in the gray matter of the brain (85 percent). Generally, the younger and more active the tissue, the greater is the concentration of water. Of the water in the human body, 37 percent is extracellular (plasma and tissue fluid), and the remaining 63 percent is intracellular. Nearly all of the water (85 percent) serves as a solvent to provide a medium for movement and interaction of ions and compounds within cells, between cells, and between the organism and the environment. The remaining water is found bound as water of hydration in proteins.

Demonstration or Laboratory Exercise:
How a Chemical Change Differs from a Physical Change

A chemical change is one in which a substance or substances is formed during a chemical reaction. In a physical change the original substances retain their chemical properties, although their size, shape, or appearance may be altered.

1. Examine a wood splint and note its appearance. Break it into two pieces and show how it is still a piece of wood in

all properties, although it has not retained its original shape. This is a physical change, as is tearing paper into smaller pieces.

2. With a pair of tongs, burn the wood splint in a flame. This is a chemical change since a new substance, ash, has been formed. The same is true when a piece of paper is burned.
3. If you pour one inch of copper sulfate solution into a test tube and then place an iron nail into the tube, after five minutes or so the nail will be coppery in color. This is a chemical change.
4. If you boil water, steam is evolved; that is a physical change. If you put a glass plate over the test tube, as the steam hits the glass plate, drops of water go back into the test tube. This is still a physical change although there was a change of state, from water to steam and back to water.

Demonstration or Laboratory Exercise:
To Show That Small Amounts of Dissolved Substances
in a Mixture Can Be Easily Separated

If you or the class did not have the time to do the basic chromatography or paper electrophoresis laboratory exercise in Unit 1, do it now, as these two processes are perfect examples of how small amounts of dissolved substances in a mixture can be separated.

Laboratory Exercise: How Components May Be Separated

Aim: To show how three different substances in a mixture can be separated.

Materials Needed: Sand, salt, iron filings, magnet, test tubes in rack, beaker, Bunsen burner, funnel, filter paper, tripod, evaporating dish, wire gauze, tongs, sheet of paper, water.

Procedure: (1) Make a mixture of sand, salt, and iron filings. Spread on a sheet of paper. (2) Pass the magnet over the mixture and iron filings will be removed. (3) Mix sand and salt with water in a beaker. Place a piece of filter paper in the funnel and place the funnel in a test tube. Pour the mixture through and the sand will remain on the filter paper. (4) Now pour solution remaining in the test tube into an evaporating dish and place the dish on the wire gauze atop the tripod. Place a slow flame under the tripod; the water will evaporate and a residue of salt will remain.

Conclusion: It is not difficult to separate any number of substances from a mixture; all that has to be done is to use the proper methods of separation for each substance in the mixture.

Laboratory Exercise: Mixtures and Compounds

Aim: To study the difference between elements, mixtures and compounds.

The universe is made up of fundamental substances called elements. The smallest part of an element that will combine with other elements is called an atom. Generally, atoms will combine with atoms to form molecules. The elements consist of molecules, the smallest part of an element to retain the properties of the element. Identical atoms are found in the molecules of elements. When two or more unlike atoms combine, a compound is formed. The molecules of the compounds keep the properties of the compound and consist of unlike atoms. There are many different substances found in the world because of the tremendous number of chemical combinations possible.

Materials Needed: Powdered sulfur, iron filings, carbon disulfide, filter paper, test tubes, test tube holder, magnet, funnel, watch glass, Bunsen burner, ring stand, ring, 250 ml beaker, balance, cork stoppers.

Procedure: (1) Take a small sheet of paper and spread 1/4 teaspoonful of iron filings on it. Bring a magnet to the filings and the magnet will be attracted. Now place 1/4 teaspoonful of sulfur on a sheet of paper and try with the magnet. It will not be attracted. (2) Into a test tube place a pinch of iron filings, and into another a pinch of sulfur. Add carbon disulfide to both to the depth of one inch. Stopper each tube and shake gently. The sulfur dissolves in the carbon disulfide while the iron filings do not. (Be certain that there are no flames near the carbon disulfide.) (3) Now mix together a little of the filings and sulfur on a sheet of paper. When you apply the magnet to it, the magnet will attract the iron and the sulfur will remain on the paper. (4) Mix the filings and the sulfur, and place into a test tube with about two inches of carbon disulfide. Stopper the tube and shake. Now filter the liquid by putting the funnel into the ring on the ring stand and filtering into a beaker. Place a few drops of the filtrate on a watch glass and let it evaporate. Within ten minutes, you will find a residue left. The iron filings will be left on the filter paper in the funnel while the sulfur crystals appear as the carbon disulfide evaporates. (5) Weigh out seven grams of iron filings and four grams of powdered sulfur and mix on a square of paper. Place into a test tube and heat until you see the two substances mixing. Cool by placing the end of the tube in cold water. The tube will break. Use the magnet on the black substance of the tube. Nothing will happen. (*Caution:* Make certain that no carbon disulfide is anywhere near an open flame.)

Conclusion: Mixtures can be made in any proportions while compounds can only be formed in definite proportions of the elements in them.

You may also say that for the iron and sulfur to unite chemically it was necessary for the mixture to be heated, and then it did not show the same characteristics as the unheated mixture. The individual elements of the mixture were recovered through filtration and evaporation.

Demonstration: The Electrolysis of Water

Although this demonstration is usually done under the unit on chemical reactions, it can be done in this unit as well to show the separation of a compound into its component elements. This demonstration will be found at the end of the next unit.

AUDIO-VISUAL MATERIALS

FILMS

Chemical Change and Temperature (E)
Chemical Changes About Us (K)
Elements, Compounds, Mixtures (K)

Explaining Matter: Atoms and Molecules (B)
Introducing Chemistry: Types of Chemical Changes (K)
Speed of Chemical Change (E)

FILMSTRIPS

Atoms and Molecules (G)
Composition of Atoms (R)
Introducing Chemical and Physical Changes (F)
Solids, Liquids, Gases, and Molecules (F)
What Things Are Made Of (F)

SINGLE-CONCEPT FILM LOOPS

Building Atom Models—Isomerism (A)
Changes of State (H)
Chemical Changes (H)
Dissolving: A Physical Change (H)
How Can You Separate a Mixture of Sugar and Starch? (L)
Mixtures and Compounds (H)
Properties of Mixtures and Compounds (Parts I and II) (I)

TRANSPARENCIES

Biochemistry (F)
Biological Chemistry and Physics (A)
Changes of State (O)
Chemistry for Biology (G)
Classes of Matter (O)
Structure of an Atom (F)

CHARTS

Biological Chemistry (F)
Matter: Properties and Changes (G)
Nature of Matter (G)

MODELS

Organic Structure Set (T)

KITS

Atomic and Molecular Construction Set (F)
Biochemistry of Living Materials Kit (A), (O), (S)
Introduction to Biochemistry Kit (A), (O), (S)
Molecular Model (A), (U)

Chemical Reactions and Bonding

UNIT 3

Since chemical reactions of all types take place in living organisms, it would be well to discuss the topic to some degree. Although atoms combine to form molecules, there are reactions between atoms and molecules. It is known that chemical reactions can be classified into several types. However, only the first three are important in biochemistry.

(1) *Synthesis:* Elements or compounds in this reaction combine to form a new compound, such as hydrogen and oxygen to water:

$$2H_2 + O_2 \rightarrow 2H_2O$$

This reaction is also called *direct combination.*

(2) *Decomposition.* This is the reverse of synthesis. A compound breaks down into its component parts such as:

$$2H_2O_2 \rightarrow 2H_2O + O_2 \quad or \quad 2H_2O \rightarrow 2H_2 + O_2$$

This reaction is also called *analysis.*

(3) *Rearrangement:* In this reaction the structure of a molecule may change without the addition or removal of any atoms.

$$CO(NH_2)_2 \rightarrow NH_4CNO$$
$$\text{urea} \qquad \text{ammonium cyanate}$$

(4) *Replacement:* Radicals or elements may exchange places. If it is a free element that changes place with a combined element, then it is called single replacement:

$$Zn + H_2SO_4 \rightarrow ZnSO_4 + H_2$$

If two compounds exchange their elements, then it is double replacement or metathesis:

$$BaCl_2 + H_2SO_4 \rightarrow BaSO_4 + 2HCl$$

(At this point use the few simple demonstrations listed and use the transparencies, filmstrips, and films as a more graphic means of illustrating the concept.)

BONDING

Covalent   Why do atoms and molecules react as they do under certain conditions? We did say that the electrons in the outer shell determine the chemical activity of an atom, since there is a natural tendency of the outer shell to fill itself with a maximum number of electrons. This can be done in either of two ways. They may mutually share their outer shell electrons, each atom thereby completing its own outer shell; or they may enter a lending-borrowing arrangement with one or more other atoms. This filling up of the outer shell process may take place between atoms of the same elements or between atoms of two or more different elements. In biological compounds, bonds formed by sharing occur more frequently than do bonds formed by lending or borrowing. The six important elements of life usually combine by sharing.

The particle formed when atoms share their electrons is called a molecule. A molecule can consist of two or more atoms, although these atoms may be of two or more different kinds. In life, the molecules are infinite in number and yet they make use of just six elements.

Molecules have the same relationship to covalent compounds as atoms do to elements. A covalent compound, therefore, is a substance composed of similar molecules.

Electrovalent   There is a second type of compound which consists of electrically charged atoms called ions, and the compound is called an ionic compound. In such a molecule there is a sharing of electrons between the oxygen and hydrogen atoms. The hydrogen atom has one electron spinning around its outer shell while the oxygen atom has six. The oxygen atom thus has an opening for two electrons to fill its outer shell, which it can do by sharing with two hydrogen atoms. Each hydrogen atom has its outer shell filled to its capacity of two, while each of the hydrogen atoms shares its lone electron with the oxygen atom, filling the oxygen's outer shell to its capacity of eight.

The diagram of this sharing becomes complicated at times, but if you would write H-O-H it would mean that the dash between the symbols stands for one pair of mutually shared electrons. In the case of CO_2 you could say O=C=O. The double line stands for two pairs of electrons being shared.

Since it becomes awkward to work out the details of each atomic structure, it is much simpler to use structural formulas in which the dash represents a pair of shared electrons. This pair of shared electrons is the chemical bond that joins the atoms.

When ionic compounds are formed as atoms join together on a lending and borrowing basis, it becomes obvious that ionic compounds are aggregates of electrically charged atoms where each individual atom must have its outer shell filled with electrons. The atoms that lend electrons will lose all the electrons in the outer shell; the outer shell disappears and an inner shell filled to capacity takes its place as the new outer shell. The atoms that borrow electrons always borrow a sufficient number to fill the outer shell. This exchange of electrons upsets the original electrical balance of the atoms involved and they are no longer electrically neutral.

When an ionic compound forms, the atoms have become different. The atom that loaned the electrons now has more protons than electrons, and since it carries an electric charge it is now a positive ion. On the other hand, the atom that borrowed the electrons has more electrons than protons and it has become a negative ion. These opposite charges attract and the result is an ionic compound. There is no direct bonding of atoms. The charged particles form large aggregates in which positively charged particles alternate with negatively charged particles. These aggregates are known as crystals.

The crystals, when dissolved in a liquid, will still maintain positively and negatively charged particles because of their strong chemical attraction for each other. These ionic compounds do not resemble the atoms of which they are composed; their properties are completely different. In your study of the biochemical process, you will find that minerals that are essential plant nutrients usually occur as ionic compounds.

Atoms may also combine to form a radical. The atoms in this particle share electrons, but the electrical charges are not completely balanced as in molecules so that the particle formed has an electrical charge and behaves as would a single ion. The radical is in reality a group of bonded atoms carrying an electrical charge. The charge can be positive, as with the ammonium radical (NH_4^+) or ion. It also can be negative, as the hydroxide radical (OH^-) or ion. When radicals combine they form ionic compounds.

(There are several good transparencies on the subject of atomic bonding which will simplify the topic for the students.)

Nonpolar Covalent Bonds   The question arises, why can two hydrogen atoms combine to form a molecule by forming a covalent bond. In the bond that links the hydrogen atoms, two electrons, one for each atom, complete the orbital. The hydrogen molecule that forms has a lower energy content and a greater stability than the two separate hydrogen atoms. The electron structure of an H_2 molecule may be represented by an electron dot diagram as follows: H:H.

Both atoms exert equal attraction for the two electrons, which means that the electron pair is shared equally between them. The electron structure and the nuclei are the same for both atoms, so that the distribution of electrical charges is equal. The resultant H_2 molecule does not have any electrical polarity. This type of bond between two identical atoms is called a nonpolar covalent bond.

Polar Covalent Bonds   In the union of hydrogen and chlorine, the chlorine exerts a stronger attraction for electrons than hydrogen. The greater electronegativity of chlorine leads to an unequal sharing of the two electrons.

The hydrogen region of the HCl molecule obtains a positive charge equal in magnitude to the negative charge of the chlorine ion, and the HCl molecule is said to be polar. The bond is a polar covalent bond, and a polar covalent bond always forms when electrons are shared by two atoms whose electronegativities are different.

Coordinate Covalent Bonds   When an atom or an ion that can accept a pair of electrons unites with a particle that can donate a pair of electrons, a coordinate covalent bond is formed between the two particles. The pair of electrons shared by the two atoms is supplied by a single atom. In an ordinary covalent bond, each atom supplies one electron to form the shared pair.

Saturated and Unsaturated Bonds   In carbon compounds, carbon atoms may be bonded to one another or to different atoms through single bonds (one pair of electrons), double bonds (two pairs of electrons), or triple bonds (three pairs of electrons).

Saturated compounds are compounds that contain molecules in which carbon atoms are bonded by a single bond, such as CCl_4.

Unsaturated compounds occur when carbon atoms are bonded by double or triple bonds.

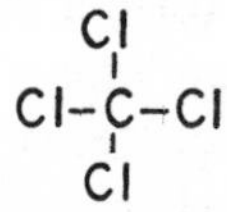

$$
\begin{array}{c}
Cl \\
| \\
Cl-C-Cl \\
| \\
Cl
\end{array}
\qquad
\overset{H}{\underset{H}{>}}C::C\overset{H}{\underset{H}{<}}
\quad \text{or} \quad
\overset{H}{\underset{H}{>}}C=C\overset{H}{\underset{H}{<}}
$$

saturated compound 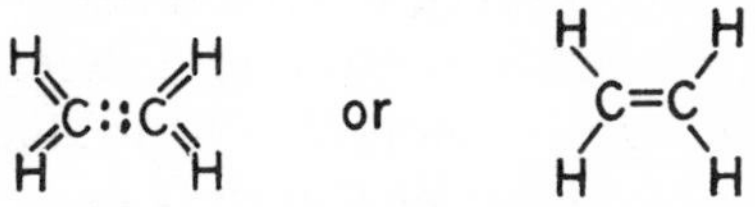   double bond

$$H:C:::C:H \quad \text{or} \quad H-C\equiv C-H$$

triple bond

Demonstration: Synthesis of Carbon Dioxide

Wrap a wire around a piece of charcoal about the size of a pea and leave about 10 cm of wire projecting for use as a handle. Heat the charcoal until it glows, then immerse it in a cylinder filled with oxygen. After the carbon has burned out, add some limewater to the cylinder and shake. The limewater becomes cloudy, since carbon dioxide has been formed directly (synthesized) from the combustion of carbon in oxygen.

Demonstration: Decomposition of Carbon Dioxide

Fill a large jar with carbon dioxide. Holding one end of a 6 cm strip of magnesium ribbon in the jaws of a pair of crucible tongs, ignite the other end with a Bunsen burner and lower it into the carbon dioxide. The magnesium continues to burn, forming a white residue. Flakes of a black substance appear on the inner walls of the jar.

There is no free oxygen in the jar. The intensity of the magnesium flame is sufficiently great to decompose the carbon dioxide; the oxygen component of the CO_2 continued to support the combustion of the magnesium while the carbon component deposits as flakes on the walls.

Demonstration: Chemical Decomposition of Water

Mix 1 ml of water with enough zinc dust to form a paste, and place it in the bottom of a Pyrex test tube. Set up apparatus as in Figure 3-1. Then place a little zinc powder in the center of the test tube by means of a spatula or a sliver of wood. Heat the paste at the bottom of the test tube with a candle flame while the dry dust in the center of the tube is being heated by a Bunsen burner flame or alcohol lamp flame.

Bubbles will appear at the end of the delivery tube and displace the water from the collecting tube. Once the tube is filled with gas, it may be lifted while still inverted and a flame applied to the open end. The gas will burn with a rather invisible flame but will produce a substantial amount of heat.

This happens between water which is composed of oxygen and hydrogen. As the water vapor from the paste passes over the zinc, much of the oxygen is absorbed by the dust as it slowly changes to zinc oxide. The residual hydrogen passed into the collecting cylinder. Hydrogen burns with a hot colorless flame.

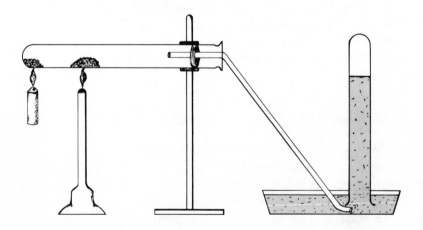

Figure 3-1 Chemical decomposition of water

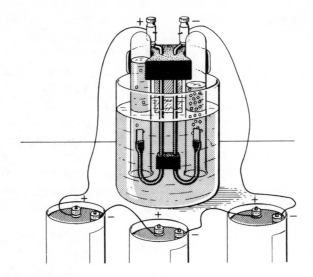

Figure 3-2 Electrolysis using the Brownlee apparatus

Demonstration: Electrolysis of Water

Energy is needed to separate a compound into its elements. In the electrolysis of water, the energy used to effect decomposition is electricity.

Fill a battery jar about one-half full of water and add a few drops of sulfuric acid. Fill two collecting tubes with the acid-water. Invert each tube under water to prevent an air space at the bottom of the tube. Lower each inverted tube over an electrode as in Figure 3-2. Connect three dry cells in series, having the positive terminals attached to the negative terminals. The negative terminal of the first cell is connected to one electrode of the apparatus. The positive terminal of the third cell is connected to the other electrode.

When the dry cells are connected to the electrodes, bubbles of gas are formed at each electrode. Allow the decomposition to proceed until one tube is almost full. Then disconnect one electrode. The ratio of the gases collected in the two tubes will be 2:1, with the largest amount of gas collected at the negative electrode.

Raise the tube filled with gas out of the apparatus, keep it inverted, and quickly place another dry tube filled with air beneath it. Hold the mouth of the tubes together firmly and invert the tubes several times so that the gases can mix. Separate the tubes, and use your thumb to seal each. Invert one and place it aside on the laboratory table. Hold the second tube at arm's length, and have a student bring a burning splint to the mouth of the tube. Release your thumb so that the gas may escape. A slight explosion occurs and flames will shoot out of the top of the tube.

Test the first tube in the same way. One of the gases that explodes when mixed with air and tested with a burning splint is hydrogen. To prove that the other tube contains oxygen, remove the other tube from the apparatus. Release any water and stopper the tube with your thumb. Hold it upright. Have a student test the gas by holding a glowing splint near the mouth of the tube as you release your thumb. The splint burns brightly. This is the laboratory test for oxygen.

AUDIO-VISUAL MATERIALS

FILMS

Chemical Bond and Atomic Structure (K)
Oxidation-Reduction (B)

FILMSTRIPS

Chemical Changes (V)
Chemical Reaction (R)
Electron Arrangement and Chemical Bonds (R)
Making Mixtures (W)
Molecules, Atoms, and Simple Reaction (R)
The Simplest Formula of a Compound (R)

SINGLE-CONCEPT FILM LOOPS

Building Atom Models (I)

TRANSPARENCIES

Atomic Bonding (F)
Chemical Bonding Set (CC)
The Covalent Bond (F)
Double Replacement Reaction (O)
Replacing an Element (O)

KITS

Organic Structure Set (F)

Solutions and Colloids

UNIT 4

From earlier studies you are aware that the living cell can be distinguished from a lifeless cell because the living cell has the ability to reproduce itself, to metabolize nutrients, to grow, to respond to stimuli, and to move. In order to maintain these characteristics, the cell must constantly take in food and excrete end products. Water, carbohydrates, lipids, proteins, vitamins, and minerals serve as food for the living cell.

The question often arises why protoplasm is made of 75-85 percent water. This high proportion of water to other substances is necessary for the solution of ions and molecules, for the suspension of molecular aggregates, and for the participation in certain chemical reactions that occur within cells.

The water content of various organisms varies from species to species, from 99.8 percent in the jellyfish to 40 percent in a very fat pig. It is interesting to note that in a given animal species, the water content is highest in the embryo and decreases with age. In a plant the seedling stage contains about 90 percent water and drops to about 50 percent at maturity.

The water content within specialized tissues in a particular animal will vary; the more active the tissue, the higher its water content. In man, the gray matter of the brain contains approximately 84 percent water, liver and muscle tissue 73-76 percent, adipose tissue 10-30 percent, and dentine of teeth only 10 percent.

The class should know that in addition to its function as a major and integral part of all living protoplasm, water serves as a medium for the transfer of food materials in the form of blood and lymph in animals, and of a nutrient solution for microorganisms. It also serves as a medium and chemical reactant in digestion, and in animals as a heat regulator and a lubricant in joints and muscles.

It is valuable to be aware of the fact that water in an organism may exist in two states: free water, which is liquid water containing the truly dissolved solutes, and serving as a dispersion medium for the colloidal particles in the protoplasm, as well as bound water. A large part of the water is bound to the colloidal particles and occurs when the water forms hydrogen bonds with proteins and other biopolymers. The bound water shows entirely different properties from that of free water.

SOLUTIONS

When you put sugar in coffee, cocoa, tea, or lemonade, or add salt to water, you know that the sugar and salt dissolve in the liquid. Solutions are very important, for if your food did not dissolve in your body, you would not be able to absorb it.

(Show the nature of solutions: solubility, temperature, etc. as demonstrations.)

At the conclusion of the demonstrations, the students will have learned what substances are able to dissolve in liquids, the rate at which substances will dissolve, and other material pertinent to a good working knowledge of the nature of solutions.

Since all substances are made up of molecules, when one substance dissolves in another the molecules of one substance mix thoroughly with the molecules of the other substance.

The students should have some recollection of the fact that a solvent is a substance that dissolves another substance, and the substance that is dissolved is called the solute. Because water dissolves so many other substances besides sugar, salt, and air, it has been called the universal solvent.

The class should also know that a physical change occurs when a solid dissolves. The result of adding a solid to a liquid to produce a clear mixture is called a solution. Solutions may be produced without an exchange of energy. However, solids usually dissolve more readily at higher temperatures.

It is possible for some substances to separate into small and even submicroscopic particles that neither dissolve nor settle to the bottom, although they are larger than molecules. Such a mixture of solid particles in a liquid is called a *suspension*. One simple way to differentiate between a suspension and a solution is to shine a flashlight beam through each. A solution will transmit the beam without reflecting any of its light—the beam will not be visible in the solution. But the particles in the suspension are large enough to reflect some of the light and thus make the beam visible.

Emulsions are suspensions of one liquid in another. Homogenized milk is an emulsion of butterfat in a complex solution. Emulsions are usually opaque. In true suspensions and emulsions, the material does not "settle out" and cannot be filtered out by ordinary means.

COLLOIDS

Colloids play an important part in biochemistry, and therefore we shall spend some time on colloidal chemistry.

Colloidal chemistry is that branch of chemistry which deals with particles of a limited range in size (6) which are larger than molecules but smaller than suspended matter which settles out on standing. Colloids are most important in living matter. Living matter is vastly complex in its composition and structure. Its basic nature is that of a mixture of extremely minute particles of solid matter suspensions in a liquid. Such suspensions are called colloidal dispersions.

The first important work with colloids was done by a Scottish scientist, Thomas Graham, in the year 1861. He added substances such as salt, sugar, starch, glue, and gelatin to water by placing them in parchment bags suspended in water. He found that while the salt and sugar readily passed through the parchment skin, the starch, glue, and gelatin were held back. Those substances which diffused easily through the parchment membrane he called crystalloids, since they form crystals. Those substances which did not pass readily through the membrane he named colloids, from the Greek word meaning "gluelike."

Graham believed that there was a fundamental difference between crystalloids and colloids, their difference in ability to diffuse through membranes. Later developments show that it was the size of the particles that determined the colloidal state rather than the nature of the substance. Since the colloidal state is a matter of particle size, almost any substance can be obtained in the colloidal state.

There is no definite boundary line of size for colloidal particles. The colloidal particles are sometimes huge molecules, as in the case of some of the proteins, while at other times they represent groups of molecules that stick to each other in clumps.

There are eight kinds of colloidal dispersions. Some of them have special names; dispersions in liquids are generally called sols. This term is most frequently used to refer to colloids formed by a solid in a liquid. The suspension of a liquid in a liquid is known as an emulsion. Mayonnaise is an emulsion formed by beating oil and egg yolk together. Colloidal dispersions in a gaseous medium are known as aerosols. Smoke is an example of an aerosol in a gas, while fog is an aerosol of a liquid in a gas.

Another special kind of colloidal system is called a gel. A gel is a jellylike material formed by the setting of a colloidal liquid. Fruit jellies are examples of gels.

If a colloidal dispersion is viewed under the ordinary microscope or seen by direct light, it appears to be transparent. If a beam of light is passed through it at right angles to the view angle, it will appear turbid because the particles scatter the light; this is known as the Tyndall effect. The size of the particles in a colloidal suspension also prevents them from passing through a parchment or an animal membrane. Thus it is possible to separate a colloidal suspension from a true solution by a process known as *dialysis*.

The relationship between the suspension and the liquid in which the particles are suspended can change from time to time. A well-known example of this property is seen when gelatin is mixed with water. Gelatin, a simple type of protein, will form what appears to be a clear solution when it is heated with water. In this condition, the colloidal particles of the gelatin are in a contracted condition and are quite widely scattered among the water molecules; in this state it is called a sol. As the gelatin cools, the particles straighten out or expand to form a firm network that traps the water within its meshes and forms a clear gel.

(Scattered throughout this piece on colloids you will find a number of demonstrations that you might do, such as the setting of gelatin, Tyndall effect, emulsification of oil. They are very simple to do and really do not need explicit instructions at the end of the unit.)

Although colloidal particles are extremely small, they expose a tremendous surface area over which chemical reactions can occur. It has been estimated that if a one-centimeter cube of a substance were divided into particles of colloidal size, its total original surface area would be increased from six square centimeters to over six million square centimeters. Since living material is colloidal in nature, this would mean that within each cell there is a vast amount of surface upon which chemical reactions can occur.

An example of this tremendous growth can be seen in the rapid absorption of water by the colloids found in a seed. If a glass is filled 1/3 full with beans or peas and the rest with water, within a few hours you will see the swollen seeds almost fill the tumbler.

In an emulsion, the homogenizing of the two liquids as in the case of oil and water breaks up the dispersed liquid (oil) into small droplets which are suspended in the other liquid. When the shaking is stopped, the droplets tend to coalesce and form large droplets which separate from the suspending medium. Such an emulsion is a temporary emulsion. The tendency to coalesce can be prevented by an emulsifying agent which surrounds the dispersed droplets with a film. When the droplets do not coalesce a permanent emulsion is formed. Milk is a familiar example of an emulsion consisting of butterfat held in suspension in water by casein, which throws a protective film around the butterfat. If the milk is homogenized by breaking down the droplets of butterfat into very small droplets, there is less tendency for the fat to separate as cream. In our digestive system, bile acts as an emulsifying agent in the digestion of fats.

The kinetic theory of gases and liquids states that molecules are in constant motion. Visible evidence of this motion in liquid was first observed by the Scottish scientist, Robert Brown, in 1827. The irregular motion shown by particles in suspension is called *Brownian* movement and is characteristic of all particles that exist in colloidal condition. The Brownian movement of colloidally suspended particles in a liquid or gas may be observed with a microscope.

Although Graham noted the difference between crystalloids and colloids by their ability to pass through or be held back by a parchment membrane, he did not understand the reason why these differences occurred. The relationship between size of particles and passage through a membrane was realized by further studies of diffusion through animal or vegetable membranes.

Such membranes allow ions and molecules of solvents and solutes to pass through their pores, but not particles in the colloidal range or larger. These membranes are called semipermeable membranes. The separation of colloids from particles in solution by use of a semipermeable membrane is known as *dialysis* (you met this term earlier).

Colloidal particles are electrically charged, some positively and some negatively. All particles in a particular system have a similar charge. The charges are obtained by the absorption of ions present in the dispersing medium. Since all particles in a particular colloidal system have a similar charge, the repulsion between the similarly charged particles keeps them from coagulating.

Many of the processes relating to living plants and animals are of a colloidal nature. The characteristics of particle size, absorption, electric charge, and dialysis are especially important in vital life processes.

Colloidal systems are important to soil fertilization. To furnish food to growing plants a good soil should contain a high proportion of very small particles, so that air and water may circulate. Particles of colloidal size absorb moisture and food for plants. Humus or decaying organic matter is added to soil because it tends to reduce clays to colloidal particles.

Many of the substances in living plants and animals are in the colloidal state. Protoplasm, the basic material of living cells, is colloidal in nature. Certain phases of digestion and blood circulation require that the substances be in colloidal form, so that some materials may pass through membranes while others are held back. If unequal passage through membranes did not exist, the body would be like a jar of homogeneous solution. A chemical reaction in one part of the body would spread rapidly to the entire body. Since life could not exist under such conditions, colloidal processes are essential to life. It would appear that protoplasm is an intricate and nonhomogeneous colloidal mixture which may shift back and forth between the sol and gel phases.

Demonstration: Solutions

(A) Solubility of Table Salt in Water

Pour 100 ml of tap water at room temperature into a beaker. Add 36 grams of table salt and stir thoroughly. The salt will dissolve at room temperature, but an additional 10 grams will remain undissolved despite violent stirring.

Various substances dissolve in water to different extents at a given temperature. When a specific volume of water will hold no more of a substance in solution, it is said to be *saturated*. At room temperature, 100 ml of water becomes saturated with 36 gm of salt. If some of the saturated solution is poured into a watch glass and allowed to evaporate slowly, the salt will gradually precipitate out in the form of transparent, cubical crystals.

(B) Water Is Not the Only Solvent

Into each of two test tubes place a small ball of camphor. Fill the first tube 3/4 full of water and the second tube 3/4 full of alcohol. Stopper each and shake each one well. Place the tubes aside for several hours and it will be seen that the camphor dissolved in the alcohol and not the water.

There are various types of solvent, and each solute has its own particular solvent. You can also show that fat will dissolve in chloroform but not water.

If you look at soda water, you will have an example of a gas in a liquid since the gas, carbon dioxide, is dissolved in the liquid, water.

(C) Stirring Dissolves Substances Faster

Into two small beakers or bottles place 50 ml of water, and into each add 5 gm of sugar. Use a stirring rod to stir one bottle, but do not touch the other. The sugar will have dissolved in the beaker that was stirred, while the other had the sugar still undissolved at the end of two minutes.

(D) The Smaller the Particles, the Faster They Dissolve

Weigh a small lump of sugar and place it into a beaker with 50 ml of water. Now weigh out an equal amount of granulated sugar and place into a beaker with 50 ml of water. The granulated sugar will dissolve much faster.

(E) Heat Will Dissolve a Substance Faster

Take two tea bags and place in separate beakers. Into one place 50 ml of cold water from the tap; into the other beaker place 50 ml of boiling water. The tea will dissolve a great deal faster in the hot water.

Along with the demonstrations, use the filmstrips and films on solutions.

Demonstration: Nature of Solutions, Colloids, Suspensions

Take four beakers of 600 ml capacity and mark them 1, 2, 3, and 4.

In beaker #1, dissolve 15 gm of table salt in 500 ml of distilled water until completely dissolved. This is your solution.

In beaker #2, add 3 gm of gelatin to 200 ml of warm distilled water and stir until it is of a homogeneous consistency. This is your colloid.

In beaker #3, add 3 gm of fine sand to 200 ml of distilled water. Stir. This is your suspension.

In beaker #4, combine 3 gm of fine clay and 200 ml of distilled water. Put it aside.

(A) Reaction to a Beam of Light

Use a strong source of light, such as a slide projector, to shine the beam of light through beakers 1, 2, and 3. The light rays will penetrate 1 and 3 but not through 2, since light can shine through solutions and suspensions but not colloids.

(B) Movement Through a Membrane

Cut four pieces of 6" lengths of dialysis tubing and soak in distilled water for thirty minutes. Tie one end of each piece of

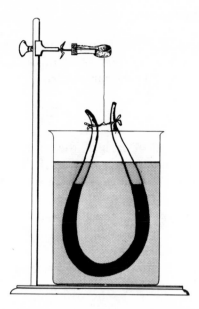

Figure 4-1 Setup with cellulose dialysis membrane

tubing with string. Pour 50 ml of the test liquid from beaker #1 into a piece of tubing. Tie the two ends together and suspend the loop from a ring stand, as illustrated in Figure 4-1. Do the same for the liquid in beaker #2 and beaker #3.

At the end of twenty minutes, with a pipette withdraw 2 ml of the distilled water in which the casing is immersed and place in a test tube. Do this for all three beakers.

When you place the liquid from the beaker with the salt solution in a test tube, a white precipitate forms, showing that there was a chloride in the water and that chloride ions can diffuse through the membrane. This occurs when you add a few drops of silver nitrate solution.

When you remove the 2 ml from the water in which the gel was and add 2 ml of concentrated nitric acid to the test tube the water turns yellow, showing that there is protein in the water. Gelatin contains protein.

No sand particles are found in the water in the beaker containing the casing for beaker #3, showing that sand did not go through the membrane. If you were to do the test with the liquid from #4, you would find that clay particles could not go through the membrane either.

(C) Testing for Suspensions

If you put a sheet of white paper under each of the original four beakers, you will notice that #1 and #2 do not settle to the bottom of the beaker while #3 and #4 do.

(D) Testing for Filtrations

If you filter 25 ml of each of the substances from the four beakers, the sand and clay will remain on the filter paper while the table salt and gelatin will not leave any residue on the filter paper.

(E) Testing for Colloids

Place 25 ml of each of the solutions in a separate evaporating dish. Heat each material to near boiling and allow to cool for thirty minutes. Only #2 will have formed a gel as the solution cooled.

If you wish, you can make up a table so that the students can immediately see the characteristics of the various liquids, solutions, colloids, and suspensions.

AUDIO-VISUAL MATERIALS

FILMS

Colloidal State (K)
Making Solutions (W)
Properties of Solution (K)
Solutions (B)

SINGLE-CONCEPT FILM LOOP

Electrolysis of Water (Y)

FILMSTRIP

Solution, Suspensions, and Colloids (V)

Acids, Bases and Salts

UNIT 5

Since water is ionized as follows: HOH into H^+ and OH^-, you realize that the number of hydrogen (H^+) ions is equal to the number of hydroxide (OH^-) ions, and the water is *neutral*.

If, on the other hand, the number (or concentration) of H^+ ions in a solution is greater than the number of OH^- ions, the solution is acidic; if the number of OH^- ions is greater than the number of H^+ ions, the solution is *basic* (or *alkaline*).

ACIDS

What do you mean when you speak of something being acidic? Anything with a sour taste is an acid. In fact, the term *acid* comes from a Latin word meaning sharp. Many familiar substances, such as sour apples, vinegar, citrus fruits, and sour milk, contain acids. Some acids are dangerously corrosive, while some are essential for life.

Point out that all acids contain hydrogen. The positive hydrogen ion is combined with a negative nonmetallic element or radical which gives the acid its name.

When acids are mixed with water, the hydrogen dissociates into hydrogen ions. These hydrogen ions are responsible for the properties of acids. Hydrogen ions are hydrates; that is, they loosely combine with water molecules to form hydronium ions,

H_3O^+. The ionization of hydrochloric acid in water can be represented by the following equation: $HCl + H_2O \rightarrow H_3O^+ + Cl^-$.

Because the water molecule which is attached to the hydrogen ion rarely participates in a reaction, the symbol H^+ for the hydrogen ions will be used. Therefore this same reaction is simply written as: $HCl \rightarrow H^+ + Cl^-$. This ionization reaction characteristic of all acids should be remembered. The more complete the ionization, the stronger the acid.

An acid reaction, such as turning blue litmus paper red, is a sign that hydrogen ions are present. Strong acids are those whose water solutions have a high concentration of hydrogen ions, such as sulfuric, nitric, and hydrochloric acids. Weak acids have few hydrogen ions in water solution, such as carbonic and phosphoric acids.

Chemists formerly defined an acid as a compound whose water solution contains hydrogen ions. This is sometimes called the Arrhenius definition of an acid. The essential activity in acid reactions is the transfer of hydrogen ions, regardless of where they come from. Since the hydronium ion, H_3O^+, in reality is a hydrated proton, we may define an acid as a substance that is a proton donor. This is called the Brönsted-Lowry definition of an acid. However, we shall use the older Arrhenius definition for our purpose.

It was previously stated that the properties of acids are due to the effect of their hydrogen ions or protons in water solution. Some of these properties are: (1) acids turn blue litmus paper red. If a strip of blue litmus is dipped in an acid solution, the blue litmus turns red. The reaction of an acid solution with blue litmus, a dye extract, is a characteristic means of identifying an acid. Other common indicators are *methyl orange,* which turns red in acid solution, and *phenolphthalein,* which is colorless in the presence of acids. (2) Acids have a sour taste. You are familiar with the sour taste of a number of substances—the taste of sour milk is due to lactic acid; vinegar is largely acetic acid; lemons, oranges, and grapefruit contain citric acid. In each case, the sour taste is caused by an acid. (3) Acids react with active metals, yielding hydrogen. (4) Acids react with bases to form salt and water. (5) Acids react with metallic oxides and carbonates. These five properties are important to remember in the field of inorganic chemistry, but numbers (1), (2), and (4) are important for biochemistry.

BASES

If you look around you, you will note that you know many of these from household use by such terms as hydroxide, caustic alkali, lye, limewater, and ammonia. Magnesium hydroxide is known as milk of magnesia, an insoluble base used medicinally as an antacid and a laxative.

According to Arrhenius, a base is a hydroxide whose water solution contains OH⁻ ions. Many substances, such as sugar and alcohol, contain the hydroxyl radical, but do not ionize in a water solution to yield the hydroxyl ion and therefore are not bases.

While many important bases are the hydroxides of metals, modern chemists define a base as a proton acceptor, since the most important thing in an acid and a base reaction is the transfer of the protons. Therefore, according to the Brönsted-Lowry theory, any substance which can accept protons (hydrogen ions) is considered to be a base. However, for present purposes, the simple concept of a base as a substance whose water solution contains the hydroxyl ion is adequate.

The characteristics of the water-soluble bases yielding the OH⁻ ion are: (1) Bases have a bitter taste and a soapy feeling. If you ever tasted such bases as the $NaOH$ present in soap lather or the $Mg(OH)_2$ in milk of magnesia you are aware of the bitter taste of bases. (2) Bases turn red litmus blue. In solutions of soluble bases the OH⁻ ion turns red litmus blue. Phenolphthalein turns purple-red in basic solutions and methyl orange becomes yellow. The change of phenolphthalein from colorless in an acid solution to red in a basic solution is useful in determining the degree of acidity or alkalinity of a water solution. (3) Bases react with acids to form salts and water. As you have already been told, when an acid and a base react they form a salt and water. This process of neutralization is essentially a reaction between the H^+ ion and the OH⁻ ion, since the metallic ion of the base and the nonmetallic ion of the acid remain unchanged.

NEUTRALIZATION

When you mix equivalent solutions of an acid and a base, the resulting solutions affect neither red nor blue litmus, because both solutions have lost their characteristic properties. Because of the transfer of hydrogen ions, which destroys the characteristic of each, acids and bases are said to neutralize each other. This process, known as neutralization, may be expressed by the equation: acid + base = salt + water + heat. In neutralization, the hydrogen ions from the acid combine with the hydroxyl ions from the base to form water: $H^+ + OH^- \rightarrow HOH$. The ions of the newly formed salt, which results from neutralization, will stay behind in solution unless the salt is insoluble. Hence these ions remain unchanged. The reaction does not reverse appreciably because water is a polar compound which is ionized to only a slight extent.

One striking phenomenon of neutralization is the release of heat. For every gram molecule of water formed, there are approximately 13,700 calories of heat evolved.

SALTS

You know the substance, salt, but what is it composed of chemically? A salt may be defined as a compound made of positive ions other than hydrogen and negative ions other than hydroxyl. In other words, a salt is an ionic compound. Soluble salts separate into their ions in solution, yielding positive metallic ions and negative nonmetallic ions.

There are a number of methods by which salts may be prepared, but for your purpose you are only concerned with the reaction of an acid and a base, where equivalent solutions of an acid and a base are mixed, and the resulting neutralization process results in a salt and water. The salt can be obtained by evaporation of the water.

(At this point, you can show the reactions of the various indicators mentioned above to acids and bases.)

HYDROGEN ION CONCENTRATION

In many reactions used in practical work, very small amounts of H^+ ions and OH^- ions make great differences in results. This is particularly true of electrolysis and fermentation reactions. In discussion of such matter you will find that concentration is very important; the concentration of hydrogen ions is designated by pH. Hence the symbol pH, followed by a number, is used to represent degrees of acidity and alkalinity.

In pure water, the H^+ ion concentration is equal to the OH^-

ion concentration. Due to the very slight degree of ionization of water, there is approximately one-ten millionth of a mole (gram-ion-weight) of each of these ions per liter. You can write the H^+ ion concentration as

$$[H^+] = \frac{1}{10,000,000} \quad or \quad 1 \times 10^{-7} \text{ grams-ions/liter}$$

If you only use the exponent (-7) and write it as a positive number, you can say that the pH of water is 7. Such a solution is neutral. Thus pH 6 means that the solution contains one part by weight of H^+ in 1,000,000 parts; pH 5, one part in 100,000. (The number in the symbol is the same as the number of ciphers.) It can be quickly seen that the concentration of H^+ ions is 10 times greater in pH 5 than in pH 6, or the relationship between pH values is in multiples of 10.

The term *pH* is defined mathematically as being equal to the logarithm of the reciprocal of the hydrogen-ion concentration, expressed in gram-ions per liter: $pH = \log \frac{1}{[H^+]}$

$$pH = \log \frac{1}{[H^+]}$$

The class should be told that solutions whose pH is more than 7 are bases; in other words, there are more hydroxyl ions than hydronium ions. A solution with a pH of 8 would be a weak base; a solution of pH 13 would be a strong base.

						Neutral							
Strong		Acid			Weak			Weak		Base			Strong
0	1	2	3	4	5	6	7	8	9	10	11	12	13	14

The pH Scale

INDICATORS

Indicators can determine the pH of a solution without any knowledge of the amounts and kinds of the acids present. Indicators are complex organic compounds that have different colors depending on whether they are in the ionized or nonionized form; that is, the color depends on the pH of the system. The pH at which an indicator changes color is determined by the chemical make-up of the indicator.

The materials responsible for the colors of most fruits and vegetables are indicators and the color changes connected with the ripening of these fruits and vegetables are a result of the general lowering of acidity. Vegetable extracts, such as litmus, and extracts of the red cabbage have been used as indicators in chemistry since the days of the alchemists.

Indicators such as litmus paper, phenolphthalein, and methyl orange are used to determine whether a solution is distinctly acidic or basic. However, the colors of these indicators do not change exactly at the neutral point, or a pH of 7. Thus the color change of litmus from red to blue occurs between a pH of about 5 to 8, or close to neutral. The color change of phenolphthalein from colorless to purple occurs between a pH of about 3 to 4, or in mildly acidic solution.

By using a combination of sensitive indicators, chemists have been able to devise a number of different universal indicators by which the pH of a solution may be ddetermined fairly accurately. Other indicators are the Gramercy Universal Indicator, which varies from 4 through 10.5, and Hydrion Paper, with a pH range of 1 to 11.

IMPORTANCE OF pH

Although you know that pH is most important in human digestion, it will be discussed in detail in the unit on animal digestion. There are several kinds of glands in the wall of the stomach. One secretes a dilute solution of hydrochloric acid, giving the gastric juice a pH of about 1 to 2. Other glands secrete a solution containing the protein pepsinogen. In the presence of hydrochloric acid, the pepsinogen is converted into the protein-digestion enzyme, pepsin. Pepsin hydrolyzes ingested proteins into peptide molecules by breaking the bonds between certain specific amino acid links in the chain.

The pancreatic juice contains trypsinogen, the inactive form of the protease trypsin. The trypsinogen is converted into trypsin by the action of the enzyme enterokinase, which is secreted by glands in the wall of the small intestine. Trypsin continues the work of pepsin by introducing water molecules between other amino acids in protein and peptide molecules. The action of pepsin ceases when the chyme is neutralized in the duodenum. Trypsin, however, works most rapidly at the pH of 8, which is then established.

pH and Urine   The kidney is made up of approximately one million nephrons, which are structural and functional units of the kidney. The distal tubule is the part of the nephron which leads into the collecting duct of the kidney. The distal tubule helps to preserve a constant pH in the blood by exchanging H^+ ions for Na^+ ions when the acidity of the blood tends to rise. In the rare instances when the blood becomes too alkaline, the distal tubule also acts to correct this. Consequently, in helping to maintain the pH of the blood within the normal limits of 7.3 and 7.4, the kidney can produce a urine whose pH drops as low as 4.5 or rises as high as 8.5.

pH in the Guard Cells   It has been found that considerable starch is stored within the guard cells during the night. Starch is a large insoluble molecule and exerts no osmotic effect. During the daylight hours, however, the amount of starch decreases, accompanied by an increase in the amount of glucose-l-phosphate:

starch + inorganic phosphate $\xrightarrow{\text{phosphorylase}}$ glucose-l-phosphate.

The enzyme that catalyzes this reaction also catalyzes the reverse reaction. The pH of the medium helps to determine which direction is favored. At pH 5, the formation of starch is promoted; at pH 8, the formation of glucose-l-phosphate is promoted.

When leaves are placed in solutions with a pH lower than 6.3, the stomata close. When they are placed in solutions with a higher pH (pH 8 seems to be optimum), the stomata open.

In living organisms a pH shift of as small as 0.2 can cause a serious impairment in the functioning of the organism. Therefore, living organisms must have some manner of protecting themselves from sudden changes in acidity. The protective mechanisms are called *buffers*. Buffers are mixtures of weak acid and its salt. A large number of different acid-salt pairs are used in chemical biological systems, as $Na_2HPO_4 + NaH_2PO_4$ *or* $NaHCO_3 + NH_3$.

Demonstration: Indicators

(1) With a glass-marking pencil, label one test tube A and the other B. Place 5 ml of 0.1N acetic acid into tube A and 5 ml of 0.1N sodium hydroxide into test tube B. (2) Dip a strip of red litmus into A. There is no color change. Now dip the red litmus

into B. The red litmus turns blue in a base. (3) Dip a strip of fresh blue litmus paper into A and the litmus turns red, while a piece of blue litmus in B remains unchanged. (4) Add 5 ml of red-cabbage juice to each of the two test tubes. To one of the test tubes add 2 to 4 drops of acetic acid from tube A. The juice turns red or pink. (5) Add 2 or 4 drops of the sodium hydroxide solution to the second tube of red-cabbage juice and the juice will turn blue. (6) Measure 5 ml of phenolphthalein into a clean test tube. It should be colorless. Add 3 drops of sodium hydroxide to the phenolphthalein and the solution turns red. Now add 3 drops of acetic acid to the tube of phenolphthalein and the solution becomes colorless again. (7) Place 5 ml brom thymol blue into a clean test tube, and with the aid of a drinking straw blow into the test tube until you notice the color changes to yellow-green, since carbon dioxide turns brom thymol blue to yellow-green. Now add 2 or 3 drops of sodium hydroxide to this tube and the original brom thymol blue is restored. (8) Check the pH of the acetic acid by using the pH Hydrion paper and checking the resulting color against the color chart. Do the same with the sodium hydroxide.

Demonstration: Neutralization

(1) Pour about 20 ml of dilute sodium hydroxide into a 100 ml beaker. Now add two or three drops of phenolphthalein solution to the liquid in the beaker. Note the change; it causes a red color to appear. Very carefully add dilute hydrochloric acid until the color begins to lighten; at this point continue to add hydrochloric acid *drop by drop* as you stir the liquid until the color just barely disappears.

When a sufficient quantity of any acid is added to a base, the base is said to be neutralized. In the neutralization process both the base and acid disappear, leaving a compound known as a salt and water. The neutralization process is complete when the indicator color just disappears, showing that both the acid and the base have been consumed fully in producing the salt and water.

Laboratory Exercise: The Measurement of pH

Aim: To observe the effect of dilution upon pH.

Materials Needed: For each group of two students: nine baby-food jars, 0.01M HCl, medicine dropper, 0.1M NaOH, stirring rod, pH Hydrion paper strips, paper toweling, pH Hydrion color chart, distilled water.

INTRODUCTION

(Caution: The acid and base furnished to you are quite dilute and are, therefore, not harmful. However, getting any into your eyes or mouth or on your skin should be avoided. In case of accident, flush at once with large amounts of water and notify your instructor.)

The general procedure in this exercise is to make a series of dilutions of an acid and of a base, and determine the pH of each dilution. This is followed by the neutralization of the acid and base.

MDF: The usual unit of measure of volume used in biological laboratories is the *milliliter* (ml) which is one-thousandth of a liter. Because it is impractical to sterilize the large number of pipettes needed for many classes to measure volume in milliliters, we shall use medicine droppers and shall invent for measuring—*the medi-cine-dropper-full,* abbreviated MDF.

Serial dilutions: In order to prepare a series of dilutions, each of which is 1/10th concentration of the previous solution, the procedure diagrammed in Figure 5-1 will be employed.

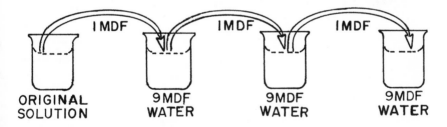

Figure 5-1 Measurement of pH

Use of the Medicine Dropper: Although the medicine drop-per seems to be a simple piece of laboratory apparatus, several precautions should be observed for its proper use. (1) The liquid in the dropper should never be allowed to run into the rubber bulb, as the liquid may be a substance which reacts with the rubber. It is also more difficult to clean the rubber bulb than the glass portion of the dropper. Always hold the medicine dropper in a vertical position, with the rubber bulb at the top. Air pressure prevents the liquid from running out of the bottom. (2) In withdrawing fluid, it sometimes makes a difference whether the rubber bulb of the medicine dropper is squeezed *before* the dropper is inserted into the liquid or *after* it is inserted. Squeezing the bulb after insertion causes agitation of the liquid as air is introduced. In some cases, such as when you are working with a protozoan culture, this agitation may be undesirable. In some cases, the introduction of air may cause oxidation of substances in the liquid. In general, do not agitate the liquid by introducing air unless there is a reason to do so. (3) Fairly uniform MDF's may be measured by squeezing the rubber bulb *tightly* each time, before insertion into the liquid. The same medicine dropper should be used for any single series of measurements. In experiments when you employ a medicine dropper to deliver a specified number of *drops*, also utilize the same medicine dropper throughout.

Procedure:

A. *Acid Dilutions*   (Wash your medicine dropper thoroughly before starting.) (1) The starting solution is hydrochloric acid at a concentration known to the chemist as 0.01M (1/100 molar). Introduce the acid into Jar A-1 to a depth of about 2.5 cm (1 inch). (2) Into each of Jars A-2, A-3, and A-4 introduce 9MDF distilled water. (3) Remove 2 MDF of solution from Jar A-1 and add it to the 9MDF in Jar A-2. Take up this diluted mixture into the medicine dropper several times in order to wash out the dropper. The concentration of acid in Jar A-2 is now 1/10 that of Jar A-1, or 0.001M HCl. (4) Continue in like manner to make serial dilutions into Jars A-3 and A-4. (5) Indicate the concentrations on the table.

B. *Basic Dilutions*   (Wash your medicine dropper thoroughly before starting.) (1) The starting solution is 0.01M NaOH (sodium hydroxide). Introduce this solution into Jar B-1 to a depth of about 2.5 cm. (2) Follow the procedure outlined above to make serial dilutions into Jars B-2, B-3, and B-4. Indicate concentrations on the table.

C. *Determination of pH*   Place a strip of pH paper on a piece of notebook paper. Using the stirring rod, transfer a drop from Jar A-1 to the strip of paper and determine the pH by comparison with the color which appears on the color chart.

Enter the pH on the table. Do the same for all the other jars. Before each determination, wash the stirring rod and dry it on paper toweling. Continue to determine and record the pH of each solution.

D. *Neutralization*   An additional jar has been furnished to you. Take 5 MDF of the original acid solution and then 5 MDF of the original basic solution and place into jar. Stir with a clean stirring rod and test the pH. It should be neutral.

E. *The pH of Biological Substances*   If you have the time, determine the pH of substances such as lemon juice, urine, blood plasma, vinegar, milk, raw egg, bicarbonate of soda, soda pop, soil, solution of bile salts, saliva, artificial gastric juice.

JAR		CONCENTRATION	pH
Acids	A-1	0.01M HCl	
	A-2		
	A-3		
	A-4		
Bases	B-1	0.01M NaOH	
	B-2		
	B-3		
	B-4		

Acknowledgments: This laboratory exercise is one that is performed at Far Rockaway High School, Far Rockaway, New York, and Mr. David Kraus, Chairman of the Science Department, has devised the ingenious method of using medicine droppers instead of pipettes and calling the measurement MDF.

AUDIO-VISUAL MATERIALS

FILMS

Acids, Bases, and Salts (K)
Indicators and pH (K)
Properties of Acids, Bases, and Salts (K)

FILMSTRIPS

Acids, Bases, and Salts (V)
Ionization and Electrolysis (R)
pH (R)

SINGLE-CONCEPT FILM LOOPS

Acid Base Indicators (A), (Z)
Acid Base Reactions in Electrolysis of Water (Y)
Buffer Solution (Y)
pH Indicators and Buffers (A), (AA)
pH Meters (BB)

TRANSPARENCIES

Neutralization (F)
pH Indicators (F)

KIT

pH Measurements and Indicator Kit (A), (CC)

Diffusion and Osmosis

UNIT 6

Would you say that diffusion is one of the most important processes occurring in living cells? Give it a little thought. Diffusion may be defined as: *the movement of molecules having the power of independent movement.* This phenomenon is exhibited in all states of matter, although movement is much more rapid in a gas than in a liquid, and is slowest in a solid.

Diffusion refers to the spread of a substance from the place of high concentration to the place of low concentration. Once the diffusion substance has spread evenly throughout the space available, equilibrium has been reached and diffusion stops, although the molecule continues to move around in the designated space.

You know that when a vial of perfume is opened in the front of a room, its odor may be detected at the back of the room within a short time due to the diffusion of the molecules through the air. Another example is a cube of sugar in a cup of tea. When a cube of sugar is dropped into the cup, the sugar will dissociate into individual molecules and go into solution throughout the liquid. Soon the sugar will become uniformly distributed throughout the tea.

Materials in living organisms have a tendency to pass from a region where they are in relatively high concentration to a region of lower concentration. This usually occurs through a cell membrane. Thus when materials are present in higher concentration outside the cells, these materials often have the ability to enter the cells by diffusion.

What accounts for this motion? The answer is Brownian movement. It is believed that all the molecules in a gas or in a solution are in constant random motion. The odds are that any object moving about at random will gradually move away from its starting point. There will be a movement of the molecules from a region of greater concentration to a region of lesser concentration. When the molecules are finally distributed evenly throughout the particular area, the process of diffusion ceases. However, the Brownian motion of the molecules continues.

In the case of diffusion between cells and their environment, there is a definite barrier involved—the cell membrane. The cell membrane is sievelike, with openings of a fairly uniform diameter. These will permit certain molecules to pass through the membranes quite freely but prevent others, the larger molecules, from doing so. A membrane of this type, showing a selective action, is called a semipermeable membrane.

Even the simplest biological membranes are semipermeable. They allow certain substances to pass through the membrane while blocking the passage of others. In general, for water-soluble substances, this choice depends mainly on molecular size. The membrane acts as though it possesses pores of a certain effective size, which permit small enough molecules to go through but block the passage of large molecules.

A second factor, added to semipermeability, is selective permeability. Cell membranes tend to pass fat-soluble molecules, almost regardless of size (there will be more about this later on). Many cell membranes tend to pass uncharged molecules much more readily than charged molecules; many pass negative ions more readily than positive ions.

A third factor regulating the penetration of substances through biological membranes is active transport. In active transport a specific mechanism exists, and work is done to carry a substance through a biological membrane. Specificity and the expenditure of energy are the earmarks of this process. Diffusion is an energy-yielding process; it can do work. Active transport is an energy-demanding process; work must be done upon it. Such active transport may take substances from a higher to a lower concentration through a membrane that would otherwise block their passage. What is even more remarkable, active transport is able to take substances from a lower to a higher concentration, that is, against the concentration gradient, bringing them to many times the concentration they possess in the medium from which they are being absorbed.

Active transport can be thought of as a process of pumping. Although little is known of the mechanism by which it occurs, it is clear that energy is required, and this is usually supplied by ATP. The specificity of the process is also apparent. Certain molecules may be passed by the membrane and concentrated, while very similar molecules are blocked. For example, a mammalian intestine absorbs galactose more rapidly than glucose, and glucose more rapidly than fructose.

Whenever a small molecule strikes the macromolecules of which the membrane is made, it rebounds. If, however, it reaches the membrane in the vicinity of a pore, it can pass through it to the other side. With a greater concentration of a given molecule on one side of the membrane than on the other, you can well imagine that there will be more collisions with the membrane on the concentrated side. There will also be more successful passages through the membrane in this direction. Although molecules do pass through the membrane in both directions, diffusion is considered to occur in the direction from higher concentration to a lower concentration.

We have said that different types of cells differ in permeability of their membranes to many substances, and similar cells may vary in this respect under different environmental conditions. However, a number of factors are responsible for the relative ease with which some substances penetrate cell membranes as compared with others. Therefore, protoplasmic membranes are selectively permeable, as in cellophane, but not entirely for the same reasons.

A selectively permeable membrane permits diffusion, but not all kinds of particles diffuse through it with the same ease. In general, smaller particles such as water molecules and glucose penetrate selectively permeable membranes more easily than do large molecules such as proteins and starches. There seem to be pores in the membrane which allow small particles to pass while preventing the large ones from passing.

The metabolism of a given organism is directly and necessarily related to the movement of materials through its cell membranes. Some of these are organic substances which are produced by the cell in excess of its own requirements, or they may be materials that are brought into the cell from an external source in order to satisfy the metabolic needs of the cell. Others are ions or inorganic compounds, and still others are materials such as water, carbon dioxide, or oxygen, which plan an important part in the chemical activities of the cell. There is a constant interchange of materials by the cell with its surrounding medium.

(There are many demonstrations and laboratory exercises described at the end of the unit.)

OSMOSIS

Since not all dissolved substances can penetrate the cell membrane with equal ease, the membrane is said to be selectively permeable. This selectivity is vital in maintaining the life of a cell. Although the cell membrane is the major structure safeguarding the cell's internal environment, other parts of the cell, such as the nucleus and mitochondrion, are also bounded by selective membranes that control their internal environment.

Whenever two different solutions are separated by a selectively permeable membrane, an osmotic system is established. Each cell, therefore, represents an osmotic unit. Osmosis may be defined as the movement of water molecules through a selectively permeable membrane. In cells, water is exchanged between the cytoplasm and the solution surrounding each cell.

When the water concentration on one side of the membrane is higher than on the other side, the water molecules will pass through the membrane and increase the volume of water on the other side. Eventually, the pressure of the water on the second side of the membrane will stop the entrance of more water. This pressure is called osmotic pressure.

However, the cell membrane is not a passive organ. While simple diffusion and osmosis account for much of the exchange of material between cells and surrounding fluids, the cell membranes also play an active part in the exchange process. Many marine algae, for instance, accumulate iodine to a concentration more than a million times greater than that of the sea. Such situations cannot be accounted for by the simple laws of diffusion or osmosis alone. Cells, therefore, have a way of forcing molecules of a particular substance to move in a direction opposite to that dictated by the laws of diffusion. This is the process of active transport.

Osmosis is but a special type of diffusion through a semi-permeable membrane. Osmosis is the basic principle involved in the process of dialysis.

Water is the common solvent for materials which are of importance to the cell. A given cell may be composed of 75 to 85 percent water and is surrounded by an aqueous medium. Water is one of only a few substances (carbon dioxide and oxygen are others) to which cell membranes are freely permeable. The exchange of water between the cell and its environment is such an important factor in cell function that it rates the special name of *osmosis.*

Whenever osmosis takes place, the solution which has the greater concentration of water molecules is called *hypotonic* while the solution with the lesser concentration of water molecules is called *hypertonic.* A solution that is equal to another in diffusible water molecules is said to be *isotonic* to it. Whenever a cell exhibits an internal pressure due to osmosis it is said to be *turgid,* and such pressure (resulting from osmosis) *turgor* pressure. This is so since the surrounding medium is more concentrated in water molecules than the protoplasmic contents of the cell. The cell does not burst because of the rigidity of the cell wall.

On the other hand, if the medium surrounding the cell is less concentrated in water molecules than the protoplasm of the cell, the water leaves a given cell. There is a loss of turgidity on the part of the cell due to osmosis. This process is called *plasmolysis* and is easily illustrated with Elodea cells. A cell whose turgidity is less than it experiences in its normal environment is said to be *flaccid.* Unless plasmolysis has occurred to a critical degree, normal turgidity may be restored by reversing the direction of osmosis.

Osmotic pressure helps move water through the conduction system in plants, and helps plant cells retain water and keep their shape and rigidity. It operates the guard cells, which open leaf stomata and permit photosynthesis to occur. In light, sugar accumulates in the guard cells, and when the water enters by osmosis the guard cells swell and separate, the stomata open, and air passes through.

In man, blood and other body fluids are kept isotonic with the body cells or else the cells may die, either by swelling or by shrinking. The red blood cells in particular are sensitive to change in osmotic pressure and depend on the kidneys and other organs of the body to regulate the body's osmotic pressure.

CELL MEMBRANES

Under an electron microscope, the membrane of a cell is seen to consist of three separate layers—the inner and outer layers are composed of proteins, with the middle layer made up of lipids. The proteins are long strandlike molecules which make it possible for the membranes to have considerable elasticity, to expand and contract according to the chemistry of the environment around the cell. The lipid layer between the two protein layers is concerned primarily with facilitating the passing of fat-solvent materials in and out of the cell.

Vital substances, such as nutritious materials, hormones, vitamins, minerals, and oxygen, pass through the cell membrane according to the needs of that particular cell, while metabolic waste products are passed out of the cell into the surrounding medium in order that they may be removed. Cell membranes are highly selective as to which molecules they will allow to enter or leave the cell.

However, on the other hand, if you were to examine a typical plant cell under the high-power microscope you would find that there is a double wall around the plant cell. The inner layer consists of a semipermeable membrane, having the two layers of protein with the lipid area between. You will also see outside of this membrane a thick wall of nonliving material which is made up of cellulose. Cellulose consists of long chains of sugar molecules which are combined to form a crystalline latticework.

A very nice cell to use as a demonstration for the semipermeability and selective permeability of living membrane is the use of the erythrocyte or red blood cell. If you take a drop of blood and suspend the cells in a very dilute sodium chloride solution (0.1% NaCl) you will notice that the red cells expand very rapidly. This swelling is known as cytolysis. The cells may swell to the point where their membranes burst, and release their hemoglobin into the solution. When this occurs, the process is referred to as hemolysis. This is due to the rapid migration of water across the cell membrane into the cytoplasm of the cells in an attempt to establish equilibrium between the more concentrated solution in the cytoplasm and the dilute saline solution in which the cells are suspended.

If the blood cells are placed in a two percent sodium chloride solution, you will find that the cells will shrink immediately as the water molecules pass from the cytoplasm into the concentrated salt solution around them. This process is plasmolysis, but since the cells develop a notched or scalloped appearance, the term *crenation* is used to describe the shrinking and shriveling of cells in a hypertonic solution.

A newly formed young plant cell will secrete sugar and gradually build up the cell wall until it will eventually become visible under the microscope. The cell membrane within the cellulose wall reacts in the same manner as that of a typical animal cell to saline solutions. If you place the cells of a long filament of spirogyra in a moderately hypertonic salt solution, you will observe plasmolysis as the cell membranes shrink away from the cellulose wall. As you add more water to the solution, the cells will once more expand and occupy the normal area within the cell wall. You now have an isotonic solution. As more water is added, a hypotonic condition will be reached and the cell membranes will expand to the very limits of the inner surface of the cellulose wall. This swelling of the cells in a very dilute or hypotonic salt solution is known as cytolysis.

Demonstration: Elodea's Salt Balance

A concentration of approximately 0.9% sodium chloride is found within an Elodea cell. You can show what will happen if this balance within the cell is upset.

(1) Take a leaf from the growing tip of the plant and place in a drop of plain water on a clean slide. Add a cover glass and review its structure with the class by using a microprojector. (2) With the help of a small piece of paper toweling, draw the water off the leaf and add a drop of 5% sodium chloride solution onto the opposite side of the cover glass. It will be drawn under the glass by the action of the filter paper. The class will see that the cytoplasm shrinks away from the cell wall. (3) If the leaf is washed and again placed in plain water, it will be seen that the cell returns to its normal condition. However, if the leaf is again placed in the salt

solution and left there for fifteen minutes, the cell dies, the cytoplasm remains in a contracted state and all streaming of the cytoplasm has stopped. Placing the cell back in plain water will not revive it. The cell membrane allowed the water molecules to enter easily and since there was salt in the water it caused the cell to act as it did.

Demonstration: Diffusion of Gases

To show how gases diffuse, open a bottle of perfume, peppermint oil, or ammonia in front of the room. Have each student stand when he first detects the odor. Those in front should be first, then the center, and finally the rear and all sides of the room.

Demonstration: Diffusion of Liquids

Half fill a beaker with molasses. Tilt the beaker and add water carefully by allowing it to run down the side only. Be certain that you do not agitate the syrup as the water is being added. Carefully place the beaker on the table and you will see two distinct layers, with the molasses gradually diffusing into the water. If you were to dip your finger into the water shortly after the demonstration was begun, you would taste the water getting slightly sweetened.

Demonstration: Diffusion of a Solid and a Liquid

Fill a beaker with water and add a pinch of copper sulfate crystals. The crystals will settle to the bottom of the beaker and soon there will be a flow of color from the crystals upward into the water. Allow the beaker to remain in sight for several days so that the students can study it every day and see the progress of the diffusion.

Demonstration: Diffusion into a Colloid, Imbibition

Select a half-dozen lima beans, fairly large and about the same size. Measure the length, width and thickness of each. Now place each one into a shallow dish of water and allow to soak overnight. Measure the beans. You will find that each of the beans is larger, the skin no longer shriveled. The bean is firm because water molecules diffused into the bean and forced the molecules of the beans apart, so that the attraction between the bean particles was decreased.

Demonstration: Osmosis

Half fill a 500 ml beaker with water. Soak a goldbeater's membrane in it for several minutes. Now hold your finger over the stem of a thistle tube and fill the bowl of the thistle tube with molasses. With a rubber band, fasten the goldbeater's membrane taut across the bowl of the tube. Invert the tube and clamp the stem to an iron stand as illustrated in Figure 6-1. Lower the thistle tube into the beaker of water so that the membrane is just beneath

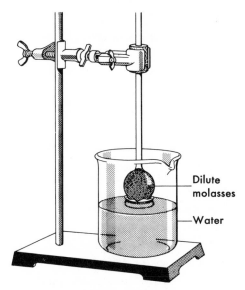

Dilute
molasses

Water

Figure 6-1 Osmosis

the surface of the water. Mark the height of the liquid in the thistle tube. Allow to stand overnight. The following day the water will have risen above the mark on the tube as the water passed from an area of greater concentration to an area of lesser concentration. The pores of the membrane allowed the water molecules to pass through, but restricted the exit of the larger sugar molecules.

Demonstration: Diffusion of Starch and Glucose

(1) Tie a knot very tightly at one end of a piece of dialysis membrane about 20 cm long. Fill the tube to within 5 cm of the top with soluble starch solution and then add about twenty drops of glucose solution to the cellophane tube. (2) Tie the top of the tube tightly with string and rinse it under running water to remove any glucose or starch that may have spilled on the outside. (3) Place the filled tube in a container with enough water to cover the solutions in the tube. Add 5 ml of iodine for each 50 ml of water in the container (Figure 6-2). (4) After fifteen minutes, test the water in the container for glucose using Fehling's or Benedict's solution. Use the water at the bottom of the container. It should test positive. (5) Meanwhile the tube will swell slightly as the water passes into it and there will be a color change to black in the bag as the iodine diffuses into the bag.

This demonstration also shows selective permeability of the membrane.

Demonstration: Diffusion Pressure

(1) Fill a 15 cm piece of dialysis tubing with glucose solution after tying a knot at one end. With a rubber band fasten the filled tube very tightly to a 1 ml pipette or a piece of glass tubing. (2) Fasten the pipette by means of a clamp to a ring stand, with the bottom of the filled tubing suspended in a container of water. Place a meter stick alongside the pipette to serve as a scale (Figure 6-3). (3) Record the position of the rising column of water measured by the meter stick at three-minute intervals. Since the water diffuses into the tubing much faster than the glucose diffuses out, there is a steady rise of the level in the tubing.

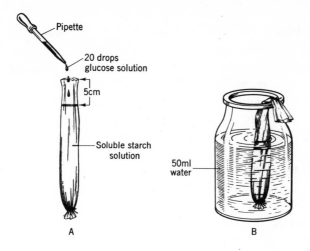

Figure 6-2 Diffusion of starch and glucose

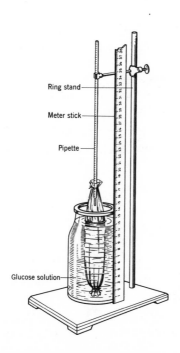

Figure 6-3 Diffusion pressure

Demonstration: Effect of an Enzyme on Diffusion

(1) Chew on a sterilized rubber band to facilitate the flow of saliva, and collect about 10 ml in a test tube. (2) Take a piece of the soaked dialysis membrane and knot at one end. Place the saliva and 10 ml of the starch solution into the membrane and place the filled bag into a small narrow-mouth bottle so that the top of the membrane can hang loosely over the top of the bottle. Pour into the bottle about 50 ml of water, or enough to cover the contents of the bag. (3) In about twenty-five minutes, take some of the contents of the bottle and put into each of two test tubes. Test one for sugar with Benedict's solution and one for starch with the iodine solution. There will be a positive test for sugar and a negative test for starch, showing that when the starch changed to sugar through the use of saliva, the sugar then diffused through the membrane.

Demonstration: Permeability of Beet Cell Membranes

(1) Cut several thin slices of raw beet, wash them, and place in distilled water. At the same time take another set of slices of the beet and boil them in distilled water for about five minutes. (2) The uncooked beets are red and the water remains clear, while the water of the boiled beets has turned reddish and the slices are paler than they were originally. This proves that living membranes are both selective and permeable, while heat destroys membranes and dead membranes lose their selectivity and thus are permeable to all molecules.

Laboratory Exercise: Diffusion

Aim: To show that molecular weight is important in diffusion.

Materials Needed: (For individual students or groups of two): 0.5% aqueous solution of malachite green (MW 365), Congo red (MW 697), and rose bengal (MW 949); a Petri dish with 10 ml ion agar, cooled; three test tubes with 10 ml ion agar, cooled; test tube rack; three medicine droppers; centimeter ruler; corkborer #9.

Procedure: (1) Prepare a 0.85% NaCl (add 2.55 gm of NaCl to 300 ml of water). Then add 2.25 gm of ion agar and bring solution to boil in a beaker, stirring continuously. You will thus have a solution of ion agar which is 0.75% in 0.85% NaCl. (2) Pour about 10 ml of the hot agar into a Petri dish to a half-inch thickness and allow to cool. (3) Pour 10 ml of the hot agar into each of the three test tubes and allow to cool. (4) With the cork borer #9, punch three holes of the same size in the agar in the Petri dish. Remove the plug from the borer each time before making another hole. Make the holes approximately the same distance apart. (5) Using a clean medicine dropper for each dye solution, place two drops of each dye solution into separate holes. At five-minute intervals examine the Petri dishes and record your observations. From your observations, what can you conclude about the relationship of the rate of diffusion of a molecule to its molecular weight? (6) Using the medicine droppers used in step five, place three drops of each solution into separate test tubes filled with the ion agar and observe the rate of diffusion of each of the solutions downward through the agar. Do you find the same results taking place in the test tubes as took place in the Petri dish?

Conclusion: Since the substance with the lowest molecular weight diffuses the fastest, you should find that in the Petri dish the ring of color around the malachite green should be the largest, the congo red second in size, and the circle around the rose bengal should be the smallest. As for the test tubes, the malachite green should have traveled down the agar farther than did the congo red, which was second, and the rose bengal traveled down the tube the least of all.

Laboratory Exercise: Diffusion

Aim: Another method to demonstrate the value of molecular weight in diffusion.

Materials Needed: Hollow glass tube about 25 mm in diameter and two feet in length, two corks to fit open ends of the tube, small

pieces of sponge, two pins, concentrated hydrochloric acid, concentrated ammonium hydroxide, ringstand, clamp to hold tube to
stand.

Procedure: (1) Place clamp in center of glass tube and attach
clamp to ringstand so that the glass tube is horizontal to the table.
(2) Place hydrochloric acid on one sponge and attach to a cork
with a pin. Place ammonium hydroxide on other sponge and
attach to other cork with pin. (3) Fit the corks into the opposite
ends of the glass tube so that no fumes escape. Watch carefully.

Observations: You will notice that the ammonium hydroxide
moves faster than the hydrochloric acid and there will be a cloudy
appearance as the ammonium hydroxide moves forward. The
hydrochloric acid is also moving, but it is moving at a slower rate
and that end of the tube will be clear. When both chemicals meet
there will be a white ring of ammonium chloride formed. The rate
of diffusion is theoretically inversely proportional to the molecular weight. The molecular weight of ammonium hydroxide is 18
while that of the hydrochloric acid is 36.5. The results will show
that the ammonium hydroxide traveled twice as far as did the
hydrochloric acid.

Laboratory Exercise: Osmosis-Diffusion

Aim: A third method of demonstrating that the molecular size of
a substance plays a role in osmosis.

Materials Needed: One piece of dialysis tubing, 8" x 1-5/8"; one
piece of dialysis tubing, 6" x 1-5/8"; string; dark Karo syrup; 400
ml beaker; #9 rubber stopper; capillary tubes about three feet in
length; ring stand with two clamps; 600 ml beaker; soluble starch
solution; IKI solution, 1/10 strength.

Procedure:

A. (1) Soak the 8" piece of dialysis tubing, then make a bag
of it by tying a knot at one end. (2) Dilute the dark Karo syrup to
half strength. Fill the bag with this solution and close it with the
rubber stopper in which the glass capillary tubing has been
inserted. (3) Make certain that the syrup completely fills the bag
after the stopper has been inserted. (4) Tie the stopper securely so
that there will be no leakage. Rinse the outer part of the bag in

clear water so that no syrup is on the outside of the bag. (5) Immerse the bag containing the sugar solution in a beaker of water, as illustrated in the demonstration on diffusion pressure. (6) Mark the level on the glass tubing with wax pencil. (7) Observe the tubing at intervals of five minutes. You will notice that the level in the tubing keeps rising and if you were to watch long enough, eventually there would be no further rise in the tubing (when the equilibrium has been reached between the water in the beaker and the solution in the bag).

B. (1) Soak the second piece of dialysis membrane. Make a small bag out of it by tying one end securely in a knot. (2) Fill the bag with soluble starch solution and then tie the upper end shut. Rinse off the bag with pure water. (3) Place a 10% IKI solution in a beaker and then immerse the bag of starch solution. (4) Observe what happens to the solution in the bag and the solution in the beaker. (5) At the beginning of the experiment the contents in the bag were clear, but within thirty minutes it had turned purple or blue-black while the color of the solution in the beaker remained a rust color. (6) Thus we see that the starch could not diffuse through the membrane since the molecules were too large, but the iodine solution was able to diffuse through and thus mixed with the contents in the bag. The contents became purple due to the combination of iodine and starch.

AUDIO-VISUAL MATERIALS

FILMS

>   Diffusion and Osmosis (K)
>   Osmosis (B)

SINGLE-CONCEPT FILM LOOPS

>   Diffusion (H)
>   Diffusion Explained (F), (W)
>   Diffusion Velocities (F), (W)
>   Molecular Motion (BB)
>   Molecular Motion—Brownian Motion (F), (W)
>   Osmosis: Dialysis (A), (AA)
>   Osmosis: Effect on Cell Membrane (A), (AA)
>   Osmosis: Movement of Water (A), (AA)
>   Osmosis: Physical Changes in Plant Tissues (A), (AA)

TRANSPARENCY

Diffusion and Osmosis (O)

KITS

Osmosis and Diffusion (A)
Permeability Kit (A)

Enzymes

UNIT 7

As a teacher of biology you know that metabolism must occur in all living cells. The process, however, is dependent upon enzyme action; therefore you will find that all living cells must contain these biochemical substances.

Enzymes make possible chemical reactions in their environments which would not occur to any appreciable degree without them. By correct definition an enzyme is an organic catalyst, and a catalyst is a substance which alters the rate of a specific chemical reaction but which remains unchanged at the conclusion of the reaction. In other words, if it were not for enzymes within the cell, many vital biochemical reactions would either not take place at all or would evolve at such a slow rate that the physiological needs of the cell would not be met.

A very interesting concept has developed in recent years, for which there is considerable evidence. It is said that enzymes are related to genes and that there is a specific gene for each individual enzyme; in fact, perhaps, enzymes may be attached to the surface of the genes. Some of this was found through studies with the mold, *Neurospora Crassa*. Biochemists have evidence, too, that some of our inherited diseases are associated with faulty enzymes or enzyme systems.

Enzymes exist in "enzyme systems" which perform stepwise series of reactions. This involves a sequence of small changes in the production of end products. These systems perform many different activities in the cell. Some enzyme systems are directly responsible for the breakdown of foodstuffs to produce energy and waste products. This type of enzyme system is called exergonic, because the final reaction of a sequence of reactions yields more energy than the total amount put into the system. On the other hand, other enzyme systems are responsible for biosynthetic or building processes of the cell. These are wholly endergonic reactions and require energy to process.

The relationship between molecular structure and function is most pronounced in those compounds which are associated with living organisms. The concept of enzyme-substrate reactions falls into this category. Enzymes are rather complex organic compounds which are vital to living systems inasmuch as they make reactions possible that would otherwise occur too slowly to be of much consequence. A substrate is any substance which is altered by enzymatic action. One of the most fundamental characteristics of enzymes is their specificity, or ability to bring about a reaction with one or only a few substrates.

All known enzymes are proteins and many of their properties depend upon this fact. Their activity depends, as do many other protein properties, on the hydrogen ion concentration of the medium. Each enzyme tends to be most active over a narrow range of hydrogen ion concentration, the "pH optimum." Enzymes are rapidly destroyed by boiling.

Most enzymes are named after the substrate by changing the root to -ase. Thus the enzyme which splits sucrose into its end products, glucose and fructose, is called sucrase. There is now a trend to saying salivary amlyase for ptyalin, gastric protease for pepsin, and so on.

Biochemists have isolated several hundred enzymes in highly purified forms and have obtained some one hundred of these in crystalline form. Each time the enzymes were found to be protein, composed of long chains of amino acids, and it is the particular sequence of these which gives the enzyme its specific characteristics as an organic catalyst. Many enzymes actually consist of two parts: (a) the protein portion, termed the apoenzyme, and (b) a nonprotein portion, called a coenzyme. The combination of the apoenzyme and the coenzyme is often referred to as the holoenzyme. Some coenzymes have been demonstrated to be vitamins, such as thiamine. It is felt by some that most, if not all the vitamins, serve as coenzymes in our bodies. Some enzymes are produced in our body in an inactive form called zymogen, which becomes active only when it comes in contact with certain metal ions or inorganic substances known as activators or kinases. An excellent example of this type of enzyme is pepsin in the stomach, which is actually produced as pepsinogen, a zymogen, which is activated by the hydrochloric acid, as kinase, in the gastric juice.

Since one of the properties of an enzyme is specificity, you can well imagine that thousands of different enzymes are needed to catalyze the multitude of chemical reactions carried out by living cells.

Chemically, since enzymes are proteins, their molecules are specialized for a catalytic function rather than an energy-yielding one. Protein molecules may be extremely long, and these long molecular chains may become folded according to specific patterns. Apparently the specificity of enzymes is due not only to the sequence of amino acids at the "specific" site of a given molecule, but also to the geometry of certain folding patterns. Enzymes vary widely in molecular weight. Simple globular enzymes may have molecular weights of as low as 20,000, while conjugated globular enzymes may have molecular weights of several millions.

If you wish to draw an analogy, a given enzyme is like a key which is specific for only one lock or type of lock. By means of special techniques it has been learned that only a small portion of a given enzyme molecule represents the specific area, just as many keys might be shaped similarly except for the "notched" area. When lock and key (substrate and enzyme) come into intimate contact, the substrate acts in a highly specific manner. In some cases the reaction yields two products, and in others the product is merely altered. In still other reactions two or more substrate molecules are combined, yielding a more complex product. The enzyme remains unchanged at the end, and may serve over and over again in such reactions.

The idea of an actual combination between enzyme and its specific substrate dates back to 1913, when Michaelis and Menten proposed the theory.

You should point out to the students that as a general rule enzymes work very rapidly, catalyzing one reaction after another. By exact methods of catalyzing it has been found that enzyme molecules are capable of catalyzing as many as 50,000 reactions within a second, unbelievable as it may seem. This accounts for the fact that a very small amount of enzyme may catalyze a large amount of substrate, and a given enzyme molecule can be used over and over again by the cell until it ultimately "wears out" and is replaced. There is a high degree of enzymatic localization within the cell according to type; certain types are found only in the mitochondria, others only within microsomes, and so on.

Enzymes are usually classified according to their function. Thus oxidase catalyzes the oxidation of certain organic compounds, and results in the oxidized form of that compound and a release of energy.

The presence of enzymes permits the cell to use its foodstuffs more efficiently. At the same time, the cell can perform only those reactions for which it has the enzymes. In general, the digestive processes of animal nutrition are extracellular. Most animals have elaborate digestive tracts or cavities into which enzymes are secreted from special glands or from the cells surrounding a given cavity. Many have a number of enzymes which work on carbohydrates, proteins and fats.

The embryonic plant within the seed cannot grow without a food source, and as yet it does not possess the ability to manufacture any. Under these conditions, it utilizes the stored but not diffusible starch which surrounds it by producing an enzyme, diastate, which is secreted into the storage material where it hydrolyzes the large starch molecules to their component glucose links. Thus the starch undergoes digestion, a term which is used in biology to describe the breakdown of nondiffusible food substances to a diffusible form.

(There are a number of demonstrations and laboratory exercises that can be used for this unit. Perhaps it might be a good idea to do some of the demonstrations, in particular, each day, with the laboratory exercises to be done at the end of the unit.)

EFFECTS OF TEMPERATURE

Temperature plays an important role in the rate at which enzyme reactions occur. The rate of most chemical reactions increases from two to three times with each rise of 10°C (18°F). Enzyme reactions obey this general rule. At 0°C (32°F), their action generally ceases, although a few plant enzymes function below this temperature. The rate of activity increases as the temperature rises until heat destroys the enzyme. The destruction of most animal enzymes occurs in the temperature range between 40° and 50°C (approximately 104°-122°F). In all animals and plants there is a rather definite temperature at which an enzyme functions most efficiently. This is called the optimum temperature for the enzyme. (Ask the class whether they know the optimum temperature enzymes in man.) In man, this temperature is approximately 37.5°C (98.6°F), as determined by a thermometer placed under the tongue.

NATURE OF THE SUBSTRATE

No one enzyme can break down all the molecules of carbohydrates, fats, and proteins which it encounters. It is necessary that there be many enzymes in the body to free the energy from different foods within the cells, decompose complex substances, build up new materials, or speed up oxidation. For example, one specific enzyme, such as ptyalin in saliva, will break up a complex molecule (starch) into a simpler substance (maltose). It requires the action of an enzyme (maltase) to reduce the maltose to a form of glucose which the body can use.

EFFECT OF pH

The action of many enzymes depends on the pH of the solution in which they react. A good example of the effect of hydrogen ion concentration is found in the action of ptyalin on starch. In the mouth, ptyalin acts in a solution that has a pH of approximately 6.8. When the food is swallowed and enters the stomach, it encounters the gastric juice, which has a pH varying between 1 and 3. This greatly increased acidity tends to stop the action of ptyalin and arrest further digestion of starch.

SUBSTRATE CONCENTRATION

Up to a certain concentration, increasing the concentration of substrate increases the rate of formation of end products. A point is finally reached where the enzyme is saturated with substrate and further increases in the concentration of the substrate will have no influence on the rate of formation of the end products.

ENZYME CONCENTRATION

If an excess of substrate is present, doubling the enzyme concentration usually doubles the rates of formation of end products only at the start of the reaction. The end products of the reaction often have an inhibitory effect on the enzyme and decrease its efficiency. Theoretically, as the concentration of the enzyme increases, a point could be reached where the substrate whose concentration remains constant is saturated with enzymes, and further increases in enzyme concentration would have no influence on the rate of formation of end products.

ACCUMULATION OF END PRODUCTS

The materials produced by the activities of enzymes are called their end products. It is not easy to demonstrate how the accumulation of end products affects human activities. However, laboratory experiments with various enzymes have indicated that the rate at which the enzyme activities process may be affected.

The zymase of yeast is a good example of this for you to remember. This is responsible for the splitting of starch into sugars. Other enzymes then convert sugar into alcohol and carbon dioxide. If the end product, alcohol, is not removed, its gradual accumulation will slow down the action of the enzymes and eventually bring the process to an end. It has been found that only 13 to 15 percent alcohol is produced before activities stop.

Not only are enzymes important for living organisms but there are several industries that make use of enzymes. In addition to the brewing industry and yeast, cheese manufacturers use enzymes of molds and bacteria, while farmers use bacterial enzymes in the preparation of silage.

Demonstration: Determination of Saccharogenic Time

Saccharogenic time is defined as the time in minutes required to hydrolyze a 0.25% starch solution at 40°C in a buffered solution (pH 6.8) containing 0.5% saliva (by volume) and 0.09M sodium chloride.

Collect 2 ml of saliva; divide and dilute 1 ml to 20 ml

distilled water. Pipette into each of two separate labeled test tubes 2 ml of each of three different standard solutions of starch, such as 2.0%, 1.0%, and 0.5%. Add to each of the six tubes 1.6 ml of the sodium chloride phosphate buffer (pH 6.8).

Pipette 0.5 ml of the diluted saliva into each of the six tubes and mix thoroughly. Place the tubes into a 35-40°C water bath, and at one minute intervals withdraw from each tube a few drops of the solution. Add these drops to a spot plate depression containing two drops of iodine-potassium iodide solution (0.001M). Near the end point (red color disappearance) perform this spot test at fifteen second intervals. Record the time required to reach the end point for each standard solution. The weaker the solution, the faster the time.

Demonstration: Enzymes

Every organism contains a number of different enzymes, one of which is catalase; catalase breaks down peroxides and liberates oxygen.

To show this, prepare seven tubes as follows: (1) 3mm cube of raw potato; (2) 3 ml of raw liver, mashed; (3) small insect, such as fruit fly, which has been killed by freezing; (4) a fruit fly which has been crushed with a stirring rod after freezing; (5) a piece of celery stalk cut across transversely; (6) 3mm cube of raw liver, to which add a little water to cover and heat until water boils, then discard water; (7) a cube of liver to which 95% alcohol was added to cover and let stand for at least five minutes. Discard the alcohol.

To each tube add enough 3% hydrogen peroxide to cover specimen and watch for escaping bubbles. The more bubbles, the faster the reaction.

Tubes 1, 2, 4, and 5 will show activity, while 3, 6, and 7 will not. Thus we can say that the enzyme is located within the organism cells, since only cut or crushed cells display the activity, while heat, as in 6 or 7, or whole organisms such as 3, will show no activity. Heat alters the chemical bonds which maintain the shape of a molecule.

If you stop to think, you will realize that 3% hydrogen peroxide poured on your skin will do nothing, but when poured on an open cut tends to fizz.

Demonstration: Effect of Varying Conditions
on Enzyme Action

Dilute some saliva 10 to 1 with distilled water, and to each of seven test tubes add 1 to 2 ml of dilute saliva. (1) To Tube 1, add five drops of starch solution and incubate at 35-50°C for ten minutes. Test for sugar should be positive. (2) Tube 2, place into boiling water and add five drops of starch solution. Allow the tube to stand for ten minutes. Test for sugar will be negative, since heat kills enzyme action. (3) Tube 3, place tube in ice and add five drops of starch solution. Allow the tube to remain in ice for ten minutes. Test for sugar will be negative, since cold kills enzyme action. (4) Tube 4, add 1 ml of pH 6 hydrochloric acid and mix. Then add five drops of starch solution. Mix and incubate at 35-40°C for ten minutes. Test for sugar will be positive. (5) Tube 5, add 1 ml of hydrochloric acid pH 3 and mix. Add five drops of starch solution and incubate at 35-40°C for ten minutes. Test for sugar will be negative because of too much acid. (6) Tube 6, add 1 ml of sodium hydroxide pH 8 and mix. Add five drops of starch solution and incubate at 35-40°C for ten minutes. Test for sugar will be positive. (7) Tube 7, add 1 ml of sodium hydroxide pH 11 and mix. Add five drops of starch solution and incubate at 35-40°C for ten minutes. Test for sugar will be negative, as it is too alkaline.

Demonstration: Cytochrome Oxidase

Cytochrome oxidase is released by the cells of living tissues. It catalyzes the union of oxygen and hydrogen. Aerobic respiration in living tissue is made possible by the secretion of the enzyme *cytochrome oxidase*. Its specific role is to transport hydrogen electrons to molecular oxygen.

(1) Cut a slice of raw potato with a clean knife. Press it firmly on a piece of filter paper that has been treated with gum guaiac. There will be a color change from brown to blue, showing areas of active cells in the potato. (2) Now wash and dry the knife carefully and slice a piece of boiled potato. Press this slice firmly on another piece of treated paper and no color change occurs.

Since heat denatures proteins, the enzyme *cytochrome oxidase* was destroyed when the potato was boiled.

Demonstration: Action of an Enzyme on Gelatin

If fresh pineapple is added to gelatin, the gelatin will not harden or gel because an enzyme from the pineapple might prevent gelling by digesting gelatin, if it is a protein, to form amino acids, which do not gel. Canned pineapple is cooked when processed; the cooking inactivates the enzyme and the gelatin will harden.

Dissolve 4 gm of gelatin in 20 ml of hot water. Divide into two portions. To one portion add 2 ml of undiluted frozen pineapple juice, and to the other portion add 2 ml of canned or boiled pineapple juice. Allow to set for ten minutes. The gelatin to which the fresh pineapple juice has been added remains soft and watery while the other has begun to gel.

Demonstration: Lipase, the Fat-Digesting Enzyme

(1) Germinate twenty-four pea seeds by soaking the peas overnight and then placing them on a wet blotter in a Petri dish for two or three days until sprouts appear. Now grind germinating seeds to pulp in water with a mortar and pestle. Filter the resulting mass and place the filtrate into a test tube. Also make a pulp of dried seeds, not germinated. (2) Dissolve over slow heat 1 gm of agar-agar in 50 ml of water. Meanwhile, mix 2.5 mg of starch with water until it forms a paste. Add the agar-agar solution and 2.0 ml of olive oil. Gently heat and stir mixture to form a smooth emulsion. Pour the emulsion into Petri dishes to form a layer 1/4 inch thick. Place in a freezer of cold refrigerator until mixture sets. (3) When set, with a medicine dropper place drops of the germinated seed extract on the surface of the layer in the Petri dish. Space the drops evenly over the surface. In the second dish place the dried pea seed extract which has not been germinated. (4) Place the dishes in incubator, preheated to 100°F, and incubate for one hour. (5) Pour a thin layer of saturated copper sulfate

solution over the surface of each incubated dish. If blue-green spots, called Copper soap, form on surface, test is positive and lipase is present.

Laboratory Exercise: Urease, an Enzyme

Materials Needed (For each student or group of two): Urease solution in dropping bottles, red litmus paper, saturated solution of urea, two test tubes, water bath (beaker), Bunsen burner, tripod, wire gauze, labels, test tube rack.

Procedure: (1)Label one test tube #1 and the other #2. Introduce about two inches of urea solution into each of the two test tubes. (2) In tube #1, fold a strip of wet red litmus paper and place in the mouth of the tube. Place the tube in the water bath. (3) In tube #2, add two drops of urease solution and shake. Place a strip of wet red litmus paper in the mouth of the tube. Keep this tube at body temperature by holding in the hand. (4) At five-minute intervals examine the litmus paper in both tubes. Note also any odor. You will notice that tube #2 has the odor of ammonia and the litmus paper has turned blue.

Urease catalyzes the breakdown of urea to carbon dioxide and ammonia. Ammonia may be detected by its odor and by its effect on wet litmus paper.

Laboratory Exercise: Enzyme Activity in Plants

Materials Needed (For each student or group of two): Germinating corn grains, prepared in advance; razor blade; iodine solution; four Petri dishes, one containing plain agar and three containing starch agar; dry corn grains; FAA (formaldehyde plus alcohol and acetic acid).

Procedure: (1) Do ahead of lab: soak corn grains and then allow to germinate on a blotter. (2) Prepare plain agar and pour into a Petri dish. (3) Prepare starch agar and pour into three Petri dishes. (4) Take several germinating corn grains and place in a dish containing FAA to kill them. Keep the grains in the solution for several hours. (5) Flood with iodine solution the Petri dish filled with plain agar and one Petri dish filled with starch agar. The dish with the starch agar will turn blue-black to indicate the test for starch. (6) Cut the germinating corn grain lengthwise in half with a razor blade and test with iodine to show that there is starch in the corn grain. (7) Place on a second Petri dish with starch agar two or three germinating corn grains which have been cut lengthwise and the cut surface placed on the agar. Allow to stand for several days. (8) Place on the third Petri dish with starch agar two or three of the killed germinating grains which have been cut lengthwise and place cut surface on the agar. Allow to stand for several days. (9) Place on the fourth Petri dish with starch agar two or three dry corn grains. (10) After several days, examine the Petri dishes with the corn grains on the agar. The Petri dish with the killed corn grains will show no change, even when the surface of the agar is flooded with iodine solution for several minutes and then poured off. The entire dish will remain blue-black. (11) Flood with the iodine the Petri dish which had the germinating corn grains. Pour off the excess iodine. Those parts where the grain was placed will be colorless, while the rest of the plate will be blue-black. We thus see that the enzymes in the grain were working on the starch and changing it to sugar.

The enzymes capable of digesting starch to sugar account for the sweet tastes of vegetables.

AUDIO-VISUAL MATERIALS

SINGLE-CONCEPT FILM LOOPS

    Bacterial Extracellular Enzymes (CC)
    Enzyme Action: Catalase (A), (AA)
    Enzyme Action: Effect of pH (A), (AA)
    Enzyme Action: Effect of Temperature (A), (AA)
    Enzyme Action: Ptyalin (A), (AA)

TRANSPARENCIES

    "Lock and Key" Theory of Enzymes (CC)
    Molecular Structure and Enzymes (CC)

Carbohydrates

UNIT 8

Did you know that carbohydrates were so named because chemists once thought of them as chains of carbon to which water molecules were attached? During the early 1800s those chemists who were studying substances such as wood, starch, and linen found that all these substances were composed mainly of carbon, hydrogen, and oxygen. The term *carbohydrate* means "water carbon," from the Greek word for water.

Carbohydrates are compounds usually made of carbon, hydrogen, and oxygen in which hydrogen and oxygen occur in the proportion of 2 to 1, so that the basic pattern is CH_2O. There are, however, many carbohydrates that contain nitrogen or sulfur.

Carbohydrates perform many vital functions in living organisms; they serve as skeletal structures in plants, insects, and crustacea, and as the outer structure of microorganisms. They are an important food reserve in the storage organs of plants, and in the liver and muscles of animals. The oxidation of carbohydrates results in most of the energy for the metabolic activities of the cells in all organisms. Humans and all animals, except those that are carnivores, derive the major portion of their food calories from the carbohydrates in their diets.

Carbohydrates are the most abundant type of organic matter found in nature, and they are found in greater quantities in plants than in animals. These two facts are easily explained. Most carbohydrates are manufactured by green plants during the process of photosynthesis. The tremendous variety of green plants found on earth accounts for the abundance of carbohydrates. However, despite their quantity, there is not a great variety of carbohydrates in living organisms. Many of them are exactly alike, regardless of where they are found—plant, animal, or human.

In general, the substances belonging to this class of compounds may be divided into three broad categories. The first of these are simple carbohydrates that contain six basic units, known as monosaccharides. They are simple sugars, very similar in structure, and are the building blocks of which large carbohydrates are composed. Examples of this carbohydrate are glucose and fructose.

The second group of carbohydrates is called disaccharides or oligosaccharides. When they are composed of two monosaccharide units united to one another through glycosidic linkage with the loss of a molecule of water, they are called disaccharides. Examples are sucrose and lactose. However, there are trisaccharides in which three monosaccharide units are united to one another. Raffinose is a trisaccharide.

The third group of carbohydrates is called polysaccharides. These represent large groups of monosaccharide units, joined through glycosidic bonds. Starch is a polysaccharide, as are cellulose and glycogen (animal starch). These are substances of large molecular weight (in the order of millions), and many contain thousands of glucose units linked together in a continuous chain.

There are a number of reactive groups present in the carbohydrate molecule which help in the identification of the specific type of carbohydrate involved. These groups are easily distinguished as the alcohol, aldehyde, and ketone groupings.

At this point, we shall go into a more detailed study of each of the classes of carbohydrates.

MONOSACCHARIDES

The empirical formula $C_6H_{12}O_6$ represents the basic structure of the carbohydrate used in human nutrition. This substance is known as the monosaccharide since it represents a single sugar group [mono-(sugar group)-saccharide]. The best known is D-glucose, found in the blood of all animals and the sap of plants. It also forms the structural unit of the most important polysaccharides.

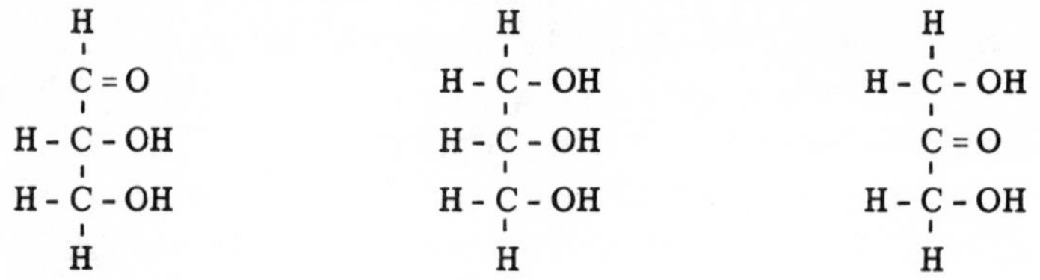

$$\begin{array}{ccc}
H & H & H \\
C=O & H-C-OH & H-C-OH \\
H-C-OH & H-C-OH & C=O \\
H-C-OH & H-C-OH & H-C-OH \\
H & H & H
\end{array}$$

Glyceraldehyde   Glycerol   Dihydroxyacetone

Figure 8-1 Simplest monosaccharides

ALPHA-RING FORM   ALDEHYDE FORM   BETA-RING FORM

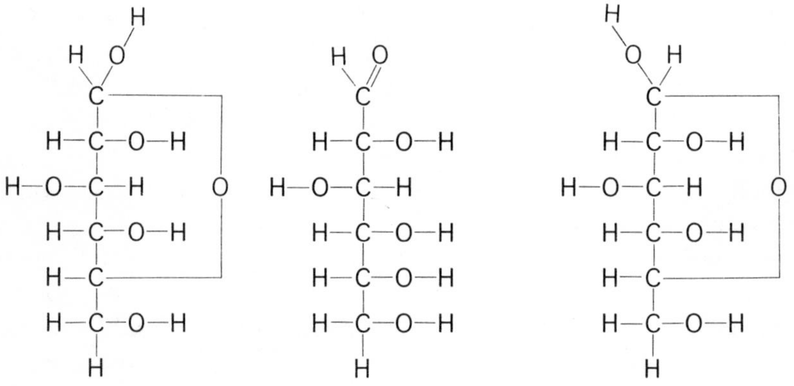

A comparison of the aldehyde form of glucose with the alpha-ring form
and the beta-ring form reveals that the two isomeric ring forms differ only
in the position of the —OH group on the first carbon atom. The compounds
of glucose differ according to the ring form involved in their formation.

Figure 8-2 Comparison of forms of glucose

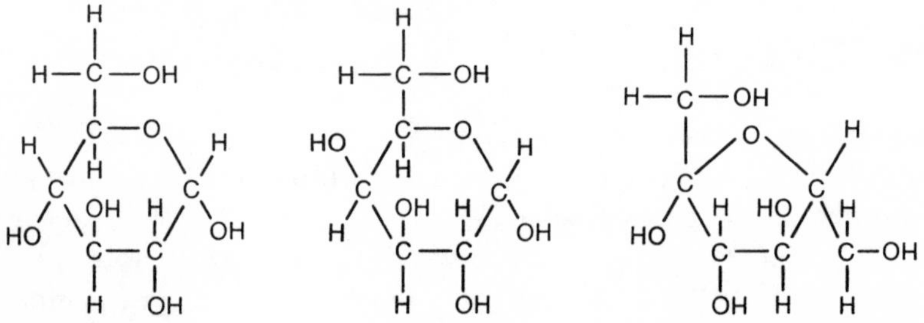

GLUCOSE   GALACTOSE   FRUCTOSE

Figure 8-3 Simple sugars

There are aldehyde or ketone derivatives of polyhydroxy-alcohols containing three or more carbon atoms; thus, glycerol may be considered the parent of the simplest monosaccharides, glyceraldehyde and dihydroxacetone. (Figure 8-1.)

Glyceraldehyde is a three-carbon sugar containing an aldehyde group. Dihydroxyacetone is a three-carbon sugar containing a ketone group. Ketone sugars are not as common in nature as aldehyde sugars, but fructose is one important exception.

In addition to the aldehyde and ketone forms, the simple sugars might form ring structures. In the cell such ring forms are very important, as many essential reactions of sugars can occur only with the ring form. (Figure 8-2.)

Since the ring form of sugars is in equilibrium with the corresponding open-chain aldehydic forms, a suitable aldehyde can react with the sugar and remove the open chain form from equilibrium. This permits the oxidation of glucose and other aldose by means of Fehling's solution. This solution is an alkaline solution of copper sulfate plus sodium potassium tartrate; the cupric ion is reduced to the cuprous form and appears as the red cuprous oxide. There are several modifications, such as Benedict's solution, which is an aqueous solution of copper sulfate, sodium citrate, and sodium carbonate. Both Fehling's and Benedict's solutions are means of identifying sugars. Both solutions are prepared by mixing a solution of copper sulfate, which is blue in color, with a strong alkali, such as sodium hydroxide. When either is heated with a solution containing a hexose, its blue color changes to yellow, red, or reddish-brown, depending on the amount of sugar present. (At this point, do the identifying tests for monosaccharides.)

Most of the simple sugars are found in plants and in plant products. Glucose is found in grapes and in honey. In fact, it is called "grape sugar" or dextrose. It is also a vital constituent of the blood of animals. Fructose, which is also called levulose or "fruit sugar," is a constituent of honey and the juice of many plants and fruits. Galactose, the other prominent monosaccharide, is rarely found as a simple sugar but as a constituent of the disaccharide, lactose, and of certain pectins and gums. (Figure 8-3.)

Glucose is important to most living organisms since the energy in the chemical bonds of the glucose molecule indirectly supplies most of the energy that most organisms use for their own activities.

(It might be a good idea to have the students taste samples of as many simple sugars as you have available. Try to include glucose and fructose, which are constituents of sucrose, a disaccharide. As an introduction to the unit on disaccharides, have the students taste sucrose also and then ask them which of the two sugars, fructose or glucose, is responsible for the sweetness of sucrose. The correct answer is that fructose is the sweetest of the sugars.)

DISACCHARIDES OR OLIGOSACCHARIDES

Disaccharides are a more complex group of sugar, as you can readily guess by just the prefix "di-." They are combinations of the ring form of sugar linked together, since the ring form is the basic sugar unit of these more complex groups.

The simplest disaccharide is formed by the splitting out of one molecule of water from the ring forms of two molecules of the same or different monosaccharides. Only three disaccharides are found free in reasonable quantities in natural products: these are sucrose, the ordinary table sugar, prepared from cane or sugar beet; lactose or milk sugar; and maltose, prepared from germinating grains. There is a fourth disaccharide, cellobiose, released by the action of microbial cellulases on cellulose; it is usually prepared by the partial acid hydrolysis of cellulose.

If you study the structural formulas of sucrose, lactose, and maltose, you will realize that sucrose is composed of glucose and fructose, lactose of galactose and glucose, while maltose is composed of two units of glucose. (Figure 8-4.)

POLYSACCHARIDES

There are two groups of polysaccharides, but both are long chains of simple sugars linked together. The skeletal or structural polysaccharides serve as rigid mechanical structure in plants and animals, and the nutrient polysaccharides act as a metabolic reserve of monosaccharides in plants and animals.

The molecules of these polysaccharides are so complex that no attempt is made to write an exact chemical formula for them. For example, starch has almost 1,000 groups. The polysaccharides

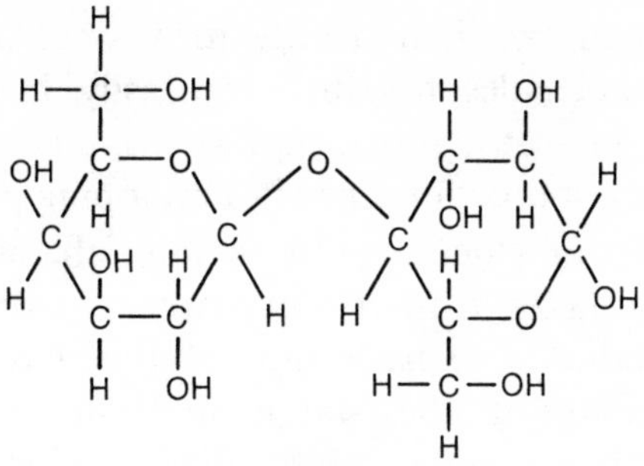

LACTOSE

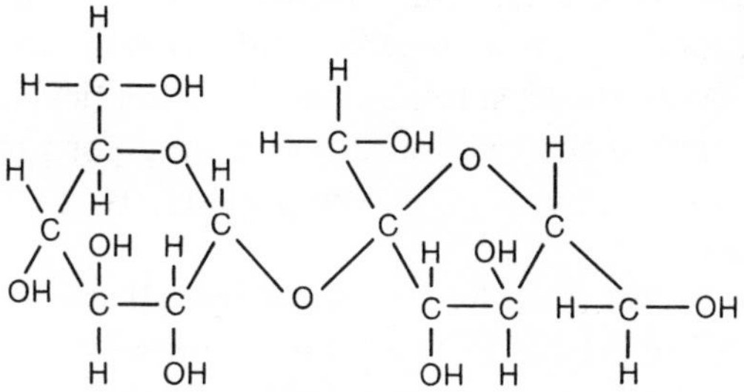

SUCROSE

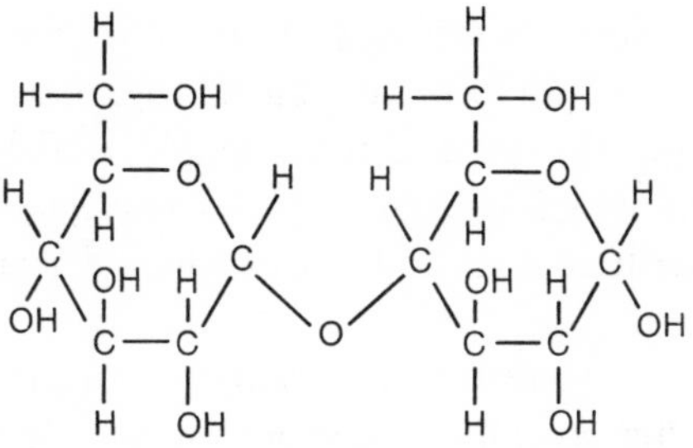

MALTOSE

Figure 8-4 Disaccharides

are therefore shown by the simplified formula $(C_6H_{10}O_5)_n$ where n stands for an unknown number.

Starches   Starches and inulin belong to the group of nutrient polysaccharides. Starches occur in the form of grains and in embryonic tissues such as potato tubers, rice, wheat, and corn seeds. Plant starch is an easily digested glucose polysaccharide that forms most of our food.

Starch is an excellent storage form since the bonds that hold the glucose molecules together are easily broken. Plant starch exists mostly in seeds, roots, and stems, where it is stored as an available energy source for a newly developing plant.

The union of glucose units in long chains produces starch, and it usually takes from twenty-two to twenty-eight units of glucose to produce one basic unit of the starch molecule. When these basic units of the starch molecule unite in a branch arrangement, they form a starch called amylopectin, which turns from red to purplish color with iodine. If, instead of branching, the union of glucose units forms a continuous straight chain, a starch called amylose results, and gives a deep-blue color with iodine. Both these starches can be classified as macromolecules and are among the macromolecules used for various structure and storage functions. Stored starch acts as an energy source for organisms.

Upon complete hydrolysis nearly all starches yield D-glucose. In the carbohydrate metabolism of animals, the important reserve polysaccharides are members of the group of substances known collectively as glycogen. Glycogen is closely related in structure to amylopectin, as it is a branching chain of glucose molecules formed in the liver and muscles of larger animals. As in starch, the bonds holding the glucose molecules together are easily broken. For this reason, the glycogen molecule makes an ideal storage molecule, from which glucose can be readily obtained. It gives a brown color with iodine and the molecule is larger in size than amylopectin.

Cellulose   Cellulose of plants is considered to be the most important of the so-called polysaccharides. It is also by far the most common.

We say that cellulose is so common and yet the student would have difficulty naming ten different objects within his immediate view that are composed largely of cellulose. The answers should include objects composed of cotton, paper, wood, linen, and all combinations of these. The students will soon realize that much of the material in the classroom, as well as much of their clothing, is made up largely of cellulose.

Cellulose, upon complete hydrolysis, will end up as D-glucose. Cellulose, however, differs in one important respect from the other polysaccharides made up of glucose molecules. The glucose units are so linked together that few organisms can split them apart with their digestive juice. Although cattle, horses, sheep, and deer eat cellulose, it is the microorganisms in their digestive tracts that split the cellulose molecules.

Cellulose differs from starch in the way the glucose units join. Starch forms alpha glucose rings, whereas cellulose forms beta glucose rings. (Alpha and beta refer to the position of the groups on the first carbon of these glucose isomers.)

Thus each glucose unit in starch has the same orientation. In cellulose, the glucose units are alternately reversed in orientation. This difference in molecular pattern results in a difference in shape as well as forming a substance that is much stronger structurally than starch. The tough units of the cellulose molecule resist the action of the digestive enzymes of many organisms, including man.

While almost all organisms can digest starch, only a few highly specialized species can digest cellulose; the protists that inhabit the intestinal tracts of termites digest cellulose. The termite makes use of the molecular fragments left by the protists. When sterile conditions are imposed on these termites and no additional protists are allowed to come in, the termites starve on a diet of cellulose. The protists also help grazing animals and some mollusks to utilize the basic units of cellulose in this way.

The major importance of cellulose lies in its use as a skeletal material by many members of the plant kingdom. It is the strength of its molecular patterns that enables cellulose to function in this respect.

Demonstration: Benedict's Solution Test for Sugar

Point out that this test is only for a simple sugar. (1) Make up a 1% solution of glucose, fructose, sucrose, and lactose, by dissolving ten grams of sugar in a liter of distilled water. (2) Place 5 ml of Benedict's solution into each of four test tubes. Add eight drops of each of the sugar solutions into a different test tube. Shake the tubes gently in order to mix the solution. (3) Place the tubes in a boiling water bath for about five to ten minutes. When the tubes are cool, the glucose has reduced so that it is deep red because it is a monosaccharide. The same will be true for the fructose. But there is no reaction in the sucrose and lactose because they are both disaccharides and will not reduce.

Demonstration: Fehling's Solution Test for Sugar

Sugars with free aldehyde or ketone groups (reducing sugars) have the property of readily reducing alkaline solutions of oxide metals such as copper, bismuth, and mercury. All the common sugars except sucrose give positive Fehling's tests. The free aldehyde or ketone group of the sugar is oxidized to a carboxyl group, while the metal is reduced to a lower valency state, and the result is a cuprous oxide which is yellow to red in color.

Place 2 ml portions of Fehling's solution (A), 7% CuSO₄ solution, in each of four test tubes. Then add 2 ml of Fehling's solution (B), 25% KOH-35% sodium potassium tartrate solution, to each of the same test tubes. Shake each tube to mix the solution. The cupric hydroxide is held in solution by the sodium potassium tartrate. Now to each tube add 1 ml of a different sugar. Mix the solution thoroughly. Place the tubes in a boiling water bath for ten minutes. The glucose and fructose will give a yellow to red color.

Demonstration: Barfoed's Test

This test distinguishes between monosaccharides and disaccharides. Monosaccharides are more reactive reducing agents and carry out the reduction in a slightly acid solution. This difference in reducing activity can be demonstrated. Barfoed's reagent contains about 7% copper acetate and 1% acetic acid.

Place 4 ml of the reagent in each of four test tubes and add 1 ml of the four original sugar solutions, one solution to each tube. Mix the solutions thoroughly and place all tubes simultaneously in a boiling water bath. Record the time necessary for the formation of a red precipitate in the bottom of each tube. The monosaccharides, glucose and fructose, will form the precipitate before the disaccharides, sucrose and lactose.

Demonstration: The Reducing Power of Glucose

Into a clean test tube pour several ml of ammonium hydroxide. Swirl the tube about to wet the sides and then pour off the residue. Add several drops of fresh ammonium hydroxide, 5 ml of silver nitrate and 5 ml of glucose solution. Shake the tube to n. the contents and place into a boiling water bath for about thre minutes. Remove tube and there should be a silver coating on the inner surface of the test tube as a result of the reduction of the ammoniacal silver oxide to the free silver form by glucose.

Demonstration: Carbohydrates and Isomers

Have two boys and two girls help you with this demonstration. The students are not to chew gum or eat candy for at least one hour prior to the demonstration.

(1) Give each student a clean sheet of paper which is marked off into six areas of equal size, and label the squares from 1 to 6. Also have ready six bottles, each containing a particular kind of sugar. One bottle contains sucrose and is so labeled. The other bottles are numbered from 2 to 6. Now have the students place an amount of sucrose sufficient to cover the surface of a dime on square #1, and place an equal amount of sugar #2 on square #2, and so on through 6. (2) Each student should moisten the tip of a finger and take up a sample of sucrose and taste it. This is the standard. Now have the students taste each of the sugars, rating each for sweetness as compared to the sucrose. The student should rinse his mouth between samples. (3) After each student has rated all the samples, tell the class that #1 was sucrose, #3 was maltose and #5 was lactose and should have tasted similarly, while #2 was glucose, #4 was galactose and #6 was fructose. When the results are tabulated, it will be found that 1, 3, and 5 tasted similar and were isomers, and 2, 4, and 6 were isomers and should have tasted similarly.

Demonstration: Carbohydrates, Yeast, and Isomers

(1) Make up a 5% solution of (a) sucrose, (b) maltose, (c) lactose, (d) glucose, (e) galactose, and (f) fructose. (2) Label six test tubes from "a" to "f". Use an amount of the sugar solution (labeled to match test tube) sufficient to cover an inverted miniature tube placed in the test tube. Then add 0.5 ml of a yeast suspension to each tube, but shake the suspension each time before withdrawing some from the tube. (3) Tip the test tube with

the small tube inside and tap it on your hand to fill the small test tube with the sugar and yeast solution. Make certain that there is no air bubble in the small tube. (4) Place the six tubes in a rack and place in an incubator at 37°C for twenty-four hours. (5) Measure the amount of gas collected in each of the small test tubes. You will find that the amount of gas varies with the sugar specimen and therefore differences in molecular structure will affect the ability of organisms to use certain chemicals. Tubes (a), (c), and (e) should have similar reactions and (b), (d), and (f) should give similar ones.

Demonstration: Conversion of Sucrose to Glucose

Boil a strong sucrose solution with a few drops of concentrated hydrochloric acid in a clean test tube for about three minutes. Test with Benedict's solution and the color will be brick red. If you tested the sucrose at the beginning of the test, you would have had no reaction with the Benedict's solution.

Acids act as a kind of "trigger" which permits sucrose to take on water and form glucose:

$$C_{12}H_{22}O_{11} + H_2O \rightarrow C_6H_{12}O_6 + C_6H_{12}O_6$$

$$\text{sucrose} \qquad \text{water} \quad \text{glucose} \qquad \text{fructose}$$

The two sugars produced have the same proportions but different properties. The difference between them is the way in which the atoms are arranged in the molecules. They are said to have different structures and are called *isomers* of each other.

Demonstration: Conversion of Starch to Glucose

(1) Mix a pinch of starch with cold water in a beaker (about half-full) and warm it gently just long enough to dissolve the starch and clear the liquid. (2) Cool the solution and divide it into four approximately equal parts in test tubes. The test tubes, numbered from one through four, should be filled about 1/3 of the way, and the remaining starch solution left in the beaker. (3) Test tube 1 when checked with iodine gives a positive test, and

test tube 2 when checked with Benedict's solution was negative. (4) Now add two drops of concentrated HCl to test tubes 3 and 4 and boil the liquid in each tube for about ten minutes. (5) Test the contents of test tube 3 with iodine and it is no longer positive; test the contents of test tube 4 with Benedict's solution and the test will be positive for glucose.

Boiling with acids causes starch to break down into glucose:

$$(C_6H_{10}O_5)_n \;+\; n(H_2O) \;\rightarrow\; n(C_6H_{12}O_6)$$
$$\text{starch} \qquad\qquad \text{water} \qquad\qquad \text{glucose}$$

Demonstration: Test for Cellulose

Use a bit of lens paper for a sample. Place it on a microscope slide and add a drop of chlor-zinc-iodine solution. The sample will turn a light violet.

Laboratory Exercise: Polysaccharides

Aim: To study the composition and means of identification of all types of polysaccharides.

Materials Needed for Polysaccharide Identification: IKI reagent (iodine dissolved in potassium iodide), 2% starch solution, 2% glycogen solution, potato starch diluted with three or four volumes of water, filter paper.

Procedure: (1) Put a drop of IKI on a piece of filter paper. It will turn violet-brown in color since there is cellulose in the paper. (2) Place 2 ml of 2% starch solution into a test tube, add a single drop of IKI reagent, and the resulting color will be a deep blue-black. If you heat the tube the color will fade, but when cooled under the tap water the color will reappear. (3) Place 2 ml of the potato starch diluted suspension in a test tube and add the IKI. The color will be blue-black, but not as deep. It, too, will fade on heating and return upon cooling under the running water. (4) Place 2 ml of 2% glycogen solution into a test tube and add a drop of IKI. The resulting color will be red-brown, which also will fade on heating and return upon cooling.

Thus you can see that different types of polysaccharides give different results with the same reagent, thus indicating that each polysaccharide has its own particular structure.

Materials Needed for Hydrolysis: Potato starch, concentrated HCl, test tubes, water bath, medicine dropper, Benedict's solution, IKI reagent, test tube rack.

Procedure: (1) Put 5 ml of potato starch into a test tube, add 5 ml of water and shake thoroughly. Then add 3 ml of concentrated HCl. Stir and place the test tube into the boiling water bath. (2) Every sixty seconds, take a drop of the solution from the tube and place into each of two test tubes. Then test each tube, one with Benedict's solution for sugar, and one with IKI for starch.

You will find a gradual change from the blue-black of unhydrolyzed starch to the red-brown of partially hydrolyzed starch and then to the absence of any reaction. As for the Benedict's test, at first it will be negative and gradually work up to a positive test when there is a negative test with IKI.

Thus you can see how the starch gradually changes into sugar in the presence of an acid.

Laboratory Exercise: Thin-Layer Chromatography of Sugars

Aim: To learn how to identify and separate various sugars out of a mixture in solution.

Materials Needed: Coated microchromatic slide (preparation in introductory unit), Coplin jars, lactose, glucose, xylose, ethyl acetate, isopropanol, water, asbestos sheet, hot plate, micro-pipette, plastic ruler in centimeters, stoppered bottles, balance, graduated cylinders, 85% phosphoric acid, naphthoresorcinol, ethyl alcohol.

Procedure: (1) Prepare as many microchromatic slides as needed. (2) Prepare 1% solution in isopropanol of lactose, glucose, and xylose. Label lactose #1, glucose #2, and xylose #3. (3) Prepare the solvent by mixing together 55 ml ethyl acetate, 36 ml isopropanol, and 12 ml of water. (4) Prepare a mixture of all sugars by mixing together equal amounts of each of the sugars into another bottle, and label it #4. (5) Number four Coplin jars from 1 to 4 and place about 2 ml of the developing solution into each of the chromatojars. Cover the jar for a few minutes so that the vapors fill the container. (6) Spot the slides by using a different sugar for each slide as well as the mixture. Thus you will have four different slides. Fill a pipette with a solution by capillary action. Touch only the liquid to the silica gel surface without breaking the surface. Arrange twenty drops adjacent to one another to form a line across the plate about 1 cm from the bottom of the slide. Allow the spots to dry thoroughly and then place Lactose #1 in the Coplin jar marked #1, and so on until you have all four slides in the proper chromatojars. (7) The solvent rises, and when it approaches the top of the silica gel layer remove each slide from the jar. All the slides should be kept in for the same period of time so that all conditions will be nearly identical. (8) Allow the slides to air dry completely and then spray with the visualizer, made up of 10 ml 85% phosphoric acid, and 0.2% naphthoresorcinol in 100 ml of ethyl alcohol. (9) When the visualizer has dried slightly, place the slides on an asbestos pad on a hot plate which is set at 200° C. (10) Within a few minutes each of the slides will begin to show color. Slide #1, lactose, will show a brownish color; slide #2, glucose, will show a reddish color; while slide #3, xylose, will show a greenish color. Slide #4, which is the mixture, will show three bands of color, and when compared to the first three slides, it can be seen that lactose is the band closest to the bottom of the slide, glucose is the middle band, and xylose is the band that traveled the farthest.

AUDIO-VISUAL MATERIALS

SINGLE-CONCEPT FILM LOOP

Biochemical Test: Sugar and Starch (A), (AA)

TRANSPARENCY

Synthesis of Carbohydrate (F)

SOLO LEARNING

Carbohydrate Structure and the Glycosidic Link (A)
Introduction to Carbohydrates (A)

Proteins

UNIT 9

From a biochemical point of view, life is characterized by its association with *proteins,* from the Greek word meaning "first rank." The only system now known for the synthesis of proteins is the living cell. Proteins are considered to be the most important group of compounds in biochemistry, since they are a major constituent of every living cell. In order to survive, all animals must have proteins in their diet, along with carbohydrates, minerals, and vitamins.

The body needs possibly 100,000 different proteins to build structures and catalyze the chemical activities of cells. Proteins are unique components of skin, hair, nails, muscle, and blood. Some hormones are proteins. The blood circulates many proteins, including antibodies which give us immunities to certain infections. Muscles contain very large amounts of protein and even teeth contain proteins. All enzymes are proteins.

Since proteins serve a diversity of functions in all living organisms, it would be well to point out the various roles of proteins. For example, the chlorophyll of green plants and the hemoglobin of the red blood cells of animals are closely related proteins. A study of the two substances will quickly show you the similarities and the differences, and it can then be pointed out to the student that there is a close relationship between plants and animals (Figure 9-1.)

The number of familiar substances that are protein in nature is unbelievable. The digestive and cellular catalysts, called

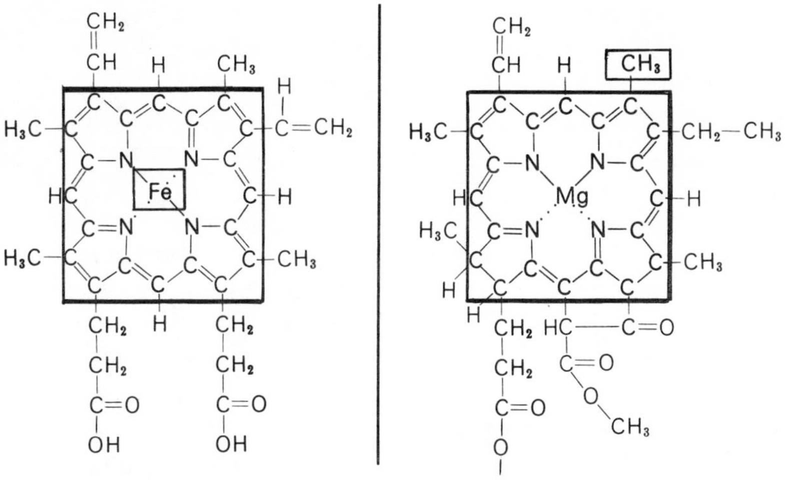

Figure 9-1 Hemoglobin-chlorophyll

enzymes, are proteins. Insulin, the hormone which regulates the body's use of sugar, is a protein. Genes, the agents which transmit inherited characteristics from parents to offspring, are giant nucleoproteins. So are viruses, the causative agents of virus diseases such as influenza and poliomyelitis. Proteins seem to be much involved in all immunities following infections. There are proteins found in food eaten by everyone, such as gelatin, the albumin of egg white, and the glutenin of wheat. How many people are aware of these proteins? To complete the roll call of proteins associated with animals and man, you must remember the myosin of muscle fibers, the collagen of bones and cartilage, as well as the keratin present in hair, horn, and feathers.

As you know, proteins are considered to be the most versatile of all chemical tools. There is no substance known in living organisms—from the smallest gas molecule to gigantic protein molecules of carbohydrates and nucleic acids—with which some protein molecules will not react. Enzymes of some organisms even react with such improbable substances as paper and leather. Proteins also have the ability to interact with other organic substances such as lipids, carbohydrates, and nucleic acids.

Proteins show extreme variations in size and shape, and their biochemical composition is more varied than most other biological materials. The smallest of the proteins have molecular weights of 6,000, while you might find others which range in size into the millions. Proteins which combine with nucleic acids are the largest

and most complex substances in cells. It is also known that
proteins have an exceedingly complex structure, undergo continuous change in living things, and are very chemically sensitive.

Normally you do not absorb whole protein molecules from
food. Enzymes help to disassemble food proteins into their amino
acid subunits. It is up to our bodies to assemble proteins from all
these amino acids, in strict accordance with the thousands of
different blueprints. Nowhere are order and precision more vital.

The slightest mistake in assembling a single protein molecule
can work grievous changes in the cell and ultimately the entire
body. Hemoglobin, which carries oxygen from lungs to tissues and
brings carbon dioxide back to the lungs, is a protein molecule with
four chains built of some 600 amino acids. A slight cellular
error—wrong assembly of just two of the 600 amino acid units—
can result in sickle cell anemia, a grave disease of red blood cells. A
number of inborn errors of metabolism, some insignificant and
some serious, are known to result from failure of an individual
person's genetic code to specify the production of a single
enzyme.

All proteins contain carbon, hydrogen, oxygen, and nitrogen,
and usually some sulfur and phosphorus. Small amounts of other
elements may be present.

(At this point it would be meaningful to demonstrate the
various components found in proteins.)

Proteins are said to contain 50 to 55% carbon, 6 to 7%
hydrogen, 20 to 23% oxygen, and 12 to 18% nitrogen. The four
major elements in protein composition are indicated by the
symbols CHON. There are thousands of various proteins which
differ in the kinds and the arrangements of the structural units of
which they are composed.

While discussing proteins, it is necessary to point out that
proteins actually are built up of substances called amino acids,
which are called "building blocks" of proteins. These amino acids
are known, but it seems as if only twenty-four are required by the
human body. Although some proteins lack one or more of these
twenty-four, most proteins are formed from all of them. The
frequency with which any amino acid occurs in a protein and the
order in which these units are linked together determine the
specific structure of the protein. Amino acids are linked together
by removal of the water molecule from between them. Cross-links

between chains are found where sulfur-containing amino acids are present. The chain is also folded.

The structure of the amino acid is:

$$\begin{array}{ccc} H & R & O \\ \backslash & | & \nearrow \\ N-C-C & \\ \diagup & | & \diagdown \\ H & H & OH \end{array}$$

The amino group is the basic end of its formula, and its formula

$$(NH_2) \text{ is: } \quad -C\overset{\displaystyle O}{\underset{\displaystyle H}{\diagup}}$$

The acidic end (carboxyl group, COOH) is at the other end. If you use the illustrations of the formulas of the various amino acids, you will notice that the simplest of the amino acids is glycine, $H_2H - CH_2 \cdot COOH$. The chemical nature of R is different for each of the different amino acids. In every case, however, the R group is relatively simple and its structural formula is known. In the amino acid there is one carbon atom called the *alpha carbon* to which the other distinct atoms or groups are bonded. Three of the latter are always the same. The alpha carbon always has one hydrogen atom, one amino group, and one acid group bonded to it. The R group varies from one hydrogen atom in the simplest amino acid, *glycine,* to large groups of atoms which may be straight chains as in *arginine,* or they may be in a ring structure, as in *tryptophan* (Figure 9-2).

(In discussing the building of proteins the chart of protein build-ups can be used, as well as other visual aids if you have them. Most biological supply firms have a good model of amino acids that can be used, or a fairly large molecular model kit. Some supply firms have a protein model that can be used to illustrate systhesis or hydrolysis.)

When proteins are assembled, either in the laboratory or in the living cell, these basic and acid groups are brought together and fused, making what is called a peptide bond (-NHCO). Thus a dipeptide contains two amino acids; a tripeptide, three amino acids; and so on. Proteins are polypeptides which contain as many as fifty or more acid units, and these may be arranged in a single chain (as in ribonuclease) or in a series of chains (as in hemoglobin and insulin.) As the carboxyl group of one amino acid combines with the basic amino group of another, water is split out. This is readily seen when you study the *protein building* formula. (Figure 9-3.)

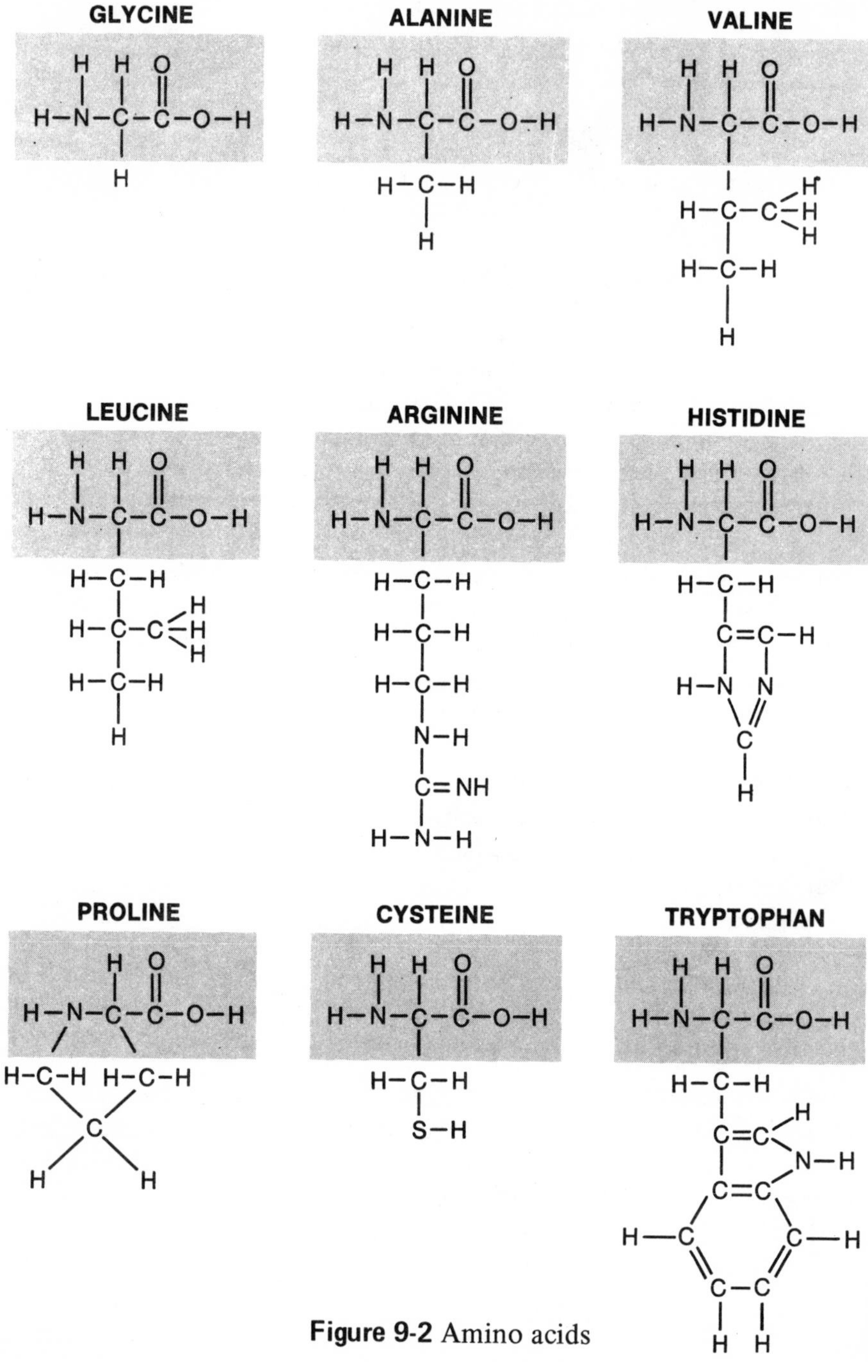

Figure 9-2 Amino acids

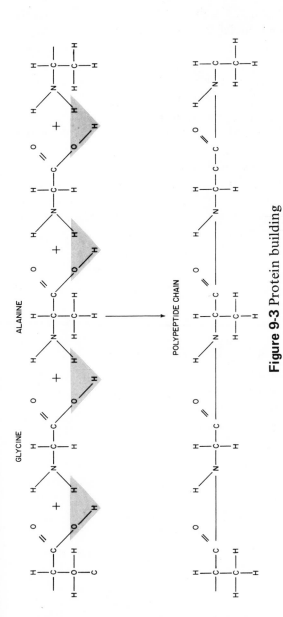

GLYCINE · **ALANINE**

POLYPEPTIDE CHAIN

Figure 9-3 Protein building

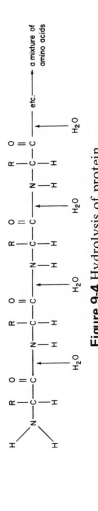

a mixture of amino acids

Figure 9-4 Hydrolysis of protein

The *primary structure* in a protein molecule is the specific sequence in which the amino acids are linked together in the molecular chain. The bonds that link one amino acid molecule with another amino acid molecule are formed between the amino group of one molecule and the acid group of another. The ability to combine with another molecule and transport it or modify it obviously depends upon the overall structure of the protein molecule. If the active or functional part were tucked away too deeply inside the molecule's coils, its chemical activity would be greatly reduced and possibly eliminated completely.

Proteins may be hydrolyzed to amino acids by heating the proteins with acids. In the digestion of proteins in the gastro-intestinal tract of higher animals, proteins are hydrolyzed to amino acids under relatively mild conditions of temperature and acidity by the proteolytic enzymes such as pepsin, trypsin, etc. (Figure 9-4.)

(There is a relatively simple laboratory exercise that can be done by the class to demonstrate the principle of hydrolysis and the gradual change of proteins to amino acids. However, it is suggested that the exercise be done after the student has seen all the protein identification tests, since this exercise is partially based on these tests.)

It was in 1902 that Emil Fischer and Franz Hofmeiter independently advanced the hypothesis that in proteins the a-amino group of one amino acid and the a-carboxyl group of another amino acid are joined, with the elimination of a molecule of water to form an amide linkage. The product of such condensation reaction is a *peptide* (Figure 9-5).

Chemists often use the terms *peptide* and *polypeptide* when describing a protein chain of less than fifty amino acids and a protein made up of over fifty amino acids.

Compounds containing peptide bonds give a characteristic purple color when treated in alkaline solution with copper sulfate. This is termed the *biuret reaction* because it is given by the substance, biuret, $NH_2CONHCONH_2$. The color deepens as the number of peptide bonds in a series of synthetic peptides is increased, and the proteins produce an especially deep violet color. The biuret reaction has provided the basis for useful methods for the determination of protein concentration in a solution. Another reaction for the detection of peptides makes use of the chlorine-

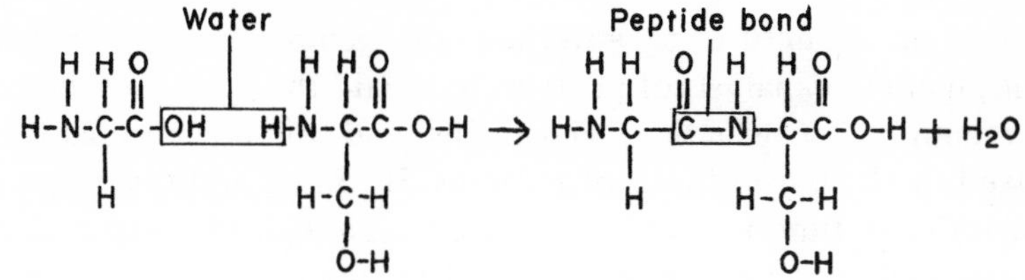

Figure 9-5 The formation of a peptide bond usually involves many enzymes. In forming the bond, a molecule of water must be released.

starch-iodine reagent. Peptides are structural intermediates between complex proteins and simple amino acids.

The vast majority of known proteins normally do not behave as long fibers, but tend to be rounded in shape; these are called *globular* or *corpuscular* proteins. The long peptide chains of each of the globular proteins must, therefore, be held in a coiled structure by chemical forces that confer upon the protein molecule its characteristic physical shape.

The coiling and folding of a protein molecule depends mostly upon the attraction between atoms or groups of atoms that occur along the length of the protein molecule. If all the amino acids were alike there might not be much variety in the shapes of protein molecules, but amino acids are not alike. The R groups on the amino acids cause each amino acid to have its own peculiar nature. It is the interaction of these R groups that influences the structure of a protein.

Many of the bonds that keep a protein molecule coiled and folded are the much weaker hydrogen bonds. This fact is important because it helps explain why many protein molecules are so sensitive. One effect of heating is an increase in the motion within the protein molecule, which may break the hydrogen bonds. Heating also may cause other kinds of bonds to break and new ones to form. This breaking of bonds is essentially what happens when a chicken's egg is heated; the egg solidifies. Heating changes the structure of the albumin molecule.

In 1954 the structural formula of the first protein was established. The protein was insulin, one of the most important hormones in our body. Its molecular formula is $C_{254}H_{377}N_{65}O_{75}S_6$.

Protein structure is governed by genes, and the recent developments in analysis of proteins have laid the basis for a better understanding of genetic mechanisms and of the mechanisms involved in the biosynthesis of proteins. It can be said that there is no significant function which can occur biologically which does not intimately involve the chemistry of proteins.

Proteins are also found outside of cells. These extracellular proteins are important supporting and strengthening materials in animals.

Green plants can make all of the amino acids from carbohydrates and from the substances that they take up from soil and water. Then, using the amino acids, they make up all of the proteins they need. On the other hand, animals cannot make up all of the amino acids they need. The human is capable of making only fifteen amino acids, and has to obtain the other nine by eating plants or by eating meat from animals that ate plants containing these other amino acids.

It is also important to realize that proteins found in different organisms are not exactly alike. Every species manufactures proteins unique to that species. Even the individuals within a species may have some protein molecules that are absolutely unique. With the exception of identical twins, no two human beings have precisely the same proteins.

There are two terms that the students should learn in connection with the chemistry of proteins; *denaturation* and *deamination*. Following is a brief resume of each term.

DENATURATION

Proteins are quite sensitive to a variety of chemical and physical agents. When exposed to these agents proteins may lose their characteristic pattern of folding, and this is denaturation. It can occur under conditions that do not affect the actual peptide linkages themselves. Nevertheless, the biochemical properties of the protein may be completely altered. Again we use the egg white as an example; the change in the white of the egg as it cooks is denaturation.

DEAMINATION

Proteins can also be used by living organisms as a source of energy. If more protein is consumed in the diet than is needed to satisfy the structural needs of the body (growth and repair), the excess can be used as fuel. After the protein is hydrolyzed into its constituent amino acids, the nitrogen-containing, or amino, portion of these molecules is removed. In man, this process is known as *deamination,* and occurs mainly in the liver. The non-nitrogen-containing residue is then oxidized in the same manner as carbohydrate and fat material. This process may even occur with the structural proteins of the body, which serve as a third reservoir of energy when needed. In times of starvation, the individual first exhausts his stores of glycogen; then the fat deposits are used, and, if starvation continues, the body proteins are then broken down and used as fuel. This means a breakdown of the structure of the body itself, and the process cannot continue indefinitely without emaciation and finally death occurring unless an outside source of food is obtained.

Demonstration: Xanthoproteic Color Test for Proteins

Place 2 ml of 2% egg albumin solution into a test tube, add 1 ml of concentrated nitric acid and heat gently. It will coagulate and turn yellow. Now cool under water and then add ammonium hydroxide to the tube. The result will be a bright orange-yellow color.

Demonstration: Millon's Color Test for Proteins

Prepare the Millon's reagent by dissolving 100 grams of mercury in 200 ml of nitric acid (specific gravity 1.42) and then dilute the solution with two volumes of distilled water. Place reagent in a dropping bottle, or buy it already prepared.

Put 2 ml of 2% egg albumin solution into a test tube and add 2 ml of Millon's reagent. Warm gently to boiling. A precipitate will form which will turn red. The color reaction is due to the hydroxyphenyl group in the protein molecule.

Demonstration: Biuret Color Test for Proteins

Prepare Biuret's reagent by adding 25 ml of 3% solution of copper sulfate per liter of 10% potassium hydroxide, or buy it already prepared.

Put 2 ml of 2% egg albumin solution into a test tube. Then add several drops of Biuret's reagent into the tube. A violet color will develop. The Biuret test is very sensitive to proteins and to polypeptides, since for proteins it runs violet and as the proteins break up into polypeptides the color test will be a pink or rose color, and finally with amino acids there will be no color change; it will remain colorless.

Demonstration: Nitroprusside Test for Proteins

The sulfhydryl groups among the proteins give a deep purple-violet color in the nitroprusside test.

To 2 ml of a 2% gelatin solution, add 1 ml of 2% sodium nitroprusside ($Na_2 Fe(CN)_5 NO$) solution, and then add two drops of 10N sodium hydroxide.

Demonstration: Ninhydrin Test for Proteins

Heat to boiling 0.5 ml of a 0.1% ninhydrin with 3 ml of an amino acid to be tested. *(Do not inhale fumes.)* The solution will turn a purplish color, the intensity of the color depending upon the concentration of the amino acid. Ninhydrin is triketo-hydrindense hydrate.

Demonstration: Nitrogen Is Present in Proteins

Place 2 ml of 2% egg albumin solution into a test tube and add an equal amount of 10% sodium hydroxide (one part of concentrated NaOH with three parts of water). Make certain that the inner rim of the tube is dry and bend a piece of red litmus paper into a U shape. Place the bottom of the U into the mouth of the tube. Heat the tube gently and as ammonia is freed from the protein solution as the protein breaks down, the litmus will turn blue.

Demonstration: Sulfur Is Present in Proteins

Place 5 ml of a 2% egg albumin solution into a test tube, and add a few drops of concentrated sulfuric acid. Moisten a piece of lead-acetate paper and crumble it into the mouth of the test tube. When the tube is heated gently the paper will turn black, showing the presence of sulfur being released.

Demonstration: Hydrolysis of Protein

As proteins are broken down into amino acids, they change first to polypeptides and then into amino acids. These changes can be demonstrated by means of the Biuret test as follows:

Number seven test tubes from 1 through 7 and add the following to each: #1—10 ml of 2% egg albumin solution, 10 ml of .5% pepsin solution; #2—10 ml of egg albumin solution, 10 ml of .5% pepsin solution and two drops of concentrated HCl; #3—10 ml of egg albumin solution, 10 ml of 0.5% pancreatin solution; #4—10 ml of egg albumin solution, 10 ml of .5% pancreatin solution and two drops of concentrated HCl; #5—10 ml of egg albumin solution; #6—10 ml of 0.5% pepsin solution; #7—10 ml of .5% pancreatin solution.

Place all seven tubes in a water bath at 40° C. Wait from thirty to forty-five minutes and then examine the tubes, adding several drops of Biuret reagent to test for protein and protein breakdown.

The results will be: #1, light violet—no hydrolysis; #2, pink—there has been change to polypeptides; #3, light violet—no hydrolysis; #4, pink—there has been a change to polypeptides; #5, violet—no change; #6, colorless as before; #7, colorless as before.

If tubes 2 and 4 are left in the water for a longer period of time, there will be a further breakdown to amino acids and the solutions in the test tubes will turn colorless upon testing with Biuret reagent.

It would be advisable to have the seven test tubes at the

beginning of the demonstration filled and ready to put into the water bath before the class comes in so that you will have the results before the class is dismissed.

Demonstration or Laboratory Exercise: Factors Affecting Protein Digestion

(1) Label six test tubes from 1 through 6. In each place 10 ml of 10% gelatin. Numbers 1 through 5 keep in a beaker of warm water to maintain the sol state, and number 6 place in a cool place to form a gel. (2) Prepare at least 6 ml of raw pineapple juice by grinding chunks of fresh pineapple in an electric blender and then straining the solid particles through fine muslin. (3) To tube #1 add 1 ml of pineapple juice, mix, let stand at room temperature twenty minutes. (4) To tube #2, add 1 ml of pineapple juice, mix, and place in water bath at 40° C for twenty minutes. (5) To tube #3, add 1 ml of pineapple juice, mix, place in a cold water bath and ice cubes for twenty minutes. (6) To tube #4, heat 1 ml of pineapple juice to boiling, add to test tube, mix, and set aside at room temperature for twenty minutes. (7) To tube #5, just keep test tube of gelatin sol at room temperature twenty minutes. (8) To tube #6, break up gel with a stirring rod into small pieces, add 1 ml of raw pineapple juice, mix with particles, let stand at room temperatures twenty minutes. (9) At the end of the twenty minutes, place all six tubes in a cool water bath for about ten minutes.

You will find that nothing has happened in #4 because the enzyme was destroyed and is no longer active. Number 5 will not change because no enzyme was added. Number 1 had the most enzyme action, while numbers 2 and 3 had less because there was a change in temperature conditions. The protein had been altered by the enzymes to different degrees, but the proteins no longer formed a colloidal gel.

Demonstration or Laboratory Exercise: Specificity of a Protein

The blood-typing laboratory exercise found in the unit on circulation can be done here to show that the identification of blood types is based on the presence of specific antigenic proteins on the surface of red blood cells.

Laboratory Exercise: Protein Analysis

Aim: To make qualitative analysis of amino acids in proteins by means of thin-layer chromatography.

We shall hydrolyze proteins into their constituent amino acids, in a method similar to the one carried out in your digestive tract, and the amino acids will then be separated through TLC.

Materials Needed: 2% egg albumin solution; phosphate buffer pH 7.0; two test tubes; pancreatic enzyme in phosphate buffer pH 7.0; 0.1% solutions of leucine, glutamic acid, aspartic acid; n-butanol; glacial acetic acid; water; ninhydrin; sprayer; asbestos sheet; hot plate; micropipettes; plastic ruler in centimeters; stoppered bottles; balance; graduated cylinders; Coplin jars; coated microchromatic slides; water bath; corks.

Procedure: (1)Into each of two test tubes place equal amounts of egg albumin solution. To one tube add 3 ml pancreatic enzyme (which was made by adding a pinch of pancreatin to 10 ml of water), plus 3 ml phosphate buffer pH7.0. Label this tube *hydrolysis.* (2) In the second tube add 3 ml of buffer alone. Label this tube *control.* Cork both. (3) Place in a water bath at 40° C until the following day. (4) Prepare microchromatic slides needed for the following day. (5) At the same time, prepare amino acid solutions of the three amino acids as follows: Prepare 0.5M HCl solution in alcohol by adding 100 ml of 1M HCl to 100 ml of 91% isopropanol. Then for each amino acid to be tested add 0.5g of the amino acid to 50 ml of the above solution. Label leucine #1, glutamic acid #2, aspartic acid #3, and the hydrolyzed substance will be #4. (6) Prepare the solvent by mixing 60 ml of n-butanol, 15 ml of acetic acid, glacial, and 25 ml of water. Number four Coplin jars and put 2 ml solvent in the botton of each, allowing solvent to permeate jars. (7) Remove the test tubes from the water bath. The tube marked *hydrolysis* will show clearing, denoting digestion of the protein. The control tube will be cloudy. (8) Spot the three amino acids plus hydrolyzed protein onto four of the coated slides. Use clean micropipettes for each solution. Air dry.

Place in the Coplin jars with the corresponding numbers of the solution and allow to migrate about twenty-five minutes. Remove the slides all at the same time and air dry. (9) Spray with the visualizer, which is prepared by adding 3 ml glacial acetic acid to 97 ml butanol. Then add 0.3g of ninhydrin to the mixture. Place mixture into sprayer and spray the slides. (10) Air dry again and then place the slides on a low-heated hot plate (200 ° C) on top of an asbestos pad and let it remain there until the colors are seen.

The slides should show as follows, in various shades of red: leucine has moved approximately 4.0cm; glutamic acid has moved approximately 3.3cm; aspartic acid has moved approximately 2.5cm. Now you can determine which is the amino that has been hydrolyzed in the egg albumin.

Demonstration or Laboratory Exercise: Electrophoresis of Amino Acid

Aim: To show that amino acids can be separated by electrophoresis.

*Materials Needed:*Electrophoresis kit, aspartic acid, leucine, lysine monohydrochloride, ninhydrin, ethanol, potassium hydroxide, oxoid barbitone acetate buffer powder, thymol, isopropanol.

Procedure: (1) Prepare each amino acid (.01M in 10% isopropanol and water). (2) Prepare a mixture combining 1.5 ml aspartic acid, 1.5 ml leucine, and 1.5 ml of lysine monohydrochloride. (3) Prepare the ninhydrin by mixing 0.2g ninhydrin in 100 ml 95% ethanol and 0.5 ml potassium hydroxide. Store in a brown stoppered bottle. (4) Prepare the oxoid barbitone acetate buffer by weighing out 6.6 gm of oxoid barbitone acetate buffer powder and adding it to 1 liter of distilled water and mix to dissolve, pH 8.6. Store in a stoppered bottle until ready to use. To preserve, add 1 ml of 5% thymol in isopropanol per liter of buffer. (5) Now follow directions as to the operation of the apparatus. Spot each amino acid and the mixture approximately 1 cm from the center of the strip at the cathode side. Set on apparatus and allow to run for fifteen minutes. (6) Remove the strips from the tanks as described. Dry the strips in a 95° C oven for five minutes. (7) Pour the ninhydrin locating reagent into a shallow glass dish to a depth of about 1/8 inch. (8) Hold the strip at each end with forceps and quickly draw the dried strip through the locating reagent. (9) Place each strip on a glass plate and hold in place with a strip of transparent tape at each end. (10) Return the strip to the 95° C oven for five minutes. (11) Observe the direction each amino acid has traveled from the point of application. The direction and rate of travel are the same whether applied as a pure substance or in a mixture containing other compounds. (12) Note the color each amino acid develops after treatment with the color-producing reagent. This color change will be similar to the one found in the TLC of amino acids, and is part of the identification of the amino acids.

AUDIO-VISUAL MATERIALS

FILMS

Chemistry of the Cell I—The Structure of Proteins and Nucleic Acids (V)

Chemistry of the Cell II—Function of DNA and RNA in Protein Synthesis (V)

FILMSTRIP

Protein Synthesis (The Living Cell Set) (A)

SINGLE-CONCEPT FILM LOOPS

Chromatograph Technique, Amino Acids (Q)

Chromatography of Amino Acids (Q)

From Amino Acids to Proteinoids (Q)

MODELS

Amino Acid Model (S), (CC)

Molecular Model Kit (S), (CC)

SOLO LEARNING

Amino Acid's Structure and the Peptide Link (A)

Introduction to Amino Acids (A)

Proteins and Deamination (A)

Lipids

UNIT 10

Together with the proteins and carbohydrates, the lipids form the bulk of organic matter of living cells. Many tissues acquire their building blocks and energy supplies in the form of substances soluble in water, such as proteins, amino acids, and sugar. However, tissues and cells also depend upon materials which are insoluble in water. This insolubility is due in part to lipids, which make up a large proportion of all membranes and subcellular particles.

Lipids are a heterogeneous collection of biochemical substances which are soluble in organic solvents such as methanol, ethanol, acetone, chloroform, ether, and benzene, and very sparingly soluble in water. Lipids are hydrophobic; most contain long-chain fatty acids. In addition, the lipid must occur in, or be a product of, living organisms.

Since lipids form membranes, you may think of them as structural components, but that is not so since they also supply energy. After a hearty meal, the tiny fat droplets are found in the blood and they are then stored in the form of fat droplets in adipose tissue cells.

Lipids are separated into groups on the basis of their chemical and physical properties. The first general group may be termed *simple lipids* or *homolipids,* composed largely of carbon and hydrogen, with relatively small amounts of oxygen. The complete oxidation of a lipid yields more than twice as many large calories of heat energy as an equal weight of starch or sugar. Upon complete hydrolysis they yield fatty acids and alcohols.

The simple lipids include the most abundant of all lipids, the fats, or triglycerides, and the less abundant waxes. The simple fats are glycerol ester of fatty acids, and fatty acids found in humans are stearic acid, palmitic acid, and oleic acid.

The simple fats are valuable in that they, unlike proteins and carbohydrates, are capable of being stored and then utilized as an energy source. Oxidation of a simple fat yields more than twice the energy yielded by an equal weight of either protein or carbohydrate.

Waxes are simple lipids with one molecule of fatty acid esterfied with one molecule of a high molecular weight, mono-hydroxyl alcohol. Waxes are found in numerous locations in animals, plants, and microorganisms, where they form a protective covering (leaves, fruits) or are present in oily secretions (animals, microorganisms). Examples of true waxes are beeswax and sperm whale wax.

The second major class of lipids are those which include the phosphorus-containing phospholipids and the galactose-containing galactolipids. Phospholipids are natural lipids which have nitrogen and phosphorus in addition to carbon, hydrogen, and oxygen. They are present in nerve tissues of animals. The most complex phospholipids are phosphatides and include lecithins and cepha-lins, which are widespread in animal and plant tissues. When one of the fatty acid groups of a simple fat is replaced by a phosphate group and a nitrogenous base, the result is a phospholipid. It has been found that phospholipids are important members of cell membranes and affect the selective permeability of those membranes (Figure 10-1).

The third major class of lipids are the derived lipids, and include the hydrolysis products of the first two classes. In addition, other compounds such as steroids, fatty aldehydes, ketones, alcohols, hydrocarbons, essential oils, fat-soluble vitamins, etc., which are produced by living cells, are included in this class.

The steroids are lipids with a ring structure. Cholesterol is a lipid and believed to be a contributing cause of the heart disease, atherosclerosis. Estrone and testosterone are hormones of the type responsible for secondary sex characteristics, female and male, respectively. It is remarkable that these two molecules differ basically by only one extra methyl group in the male hormone.

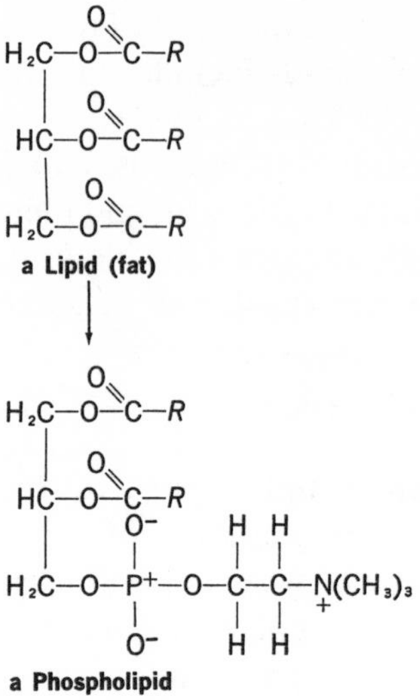

$$
\begin{array}{c}
\text{H}_2\text{C}-\text{O}-\overset{\displaystyle O}{\overset{\|}{\text{C}}}-R \\
| \quad\quad\overset{\displaystyle O}{} \\
\text{HC}-\text{O}-\overset{\displaystyle O}{\overset{\|}{\text{C}}}-R \\
| \quad\quad\overset{\displaystyle O}{} \\
\text{H}_2\text{C}-\text{O}-\overset{\displaystyle O}{\overset{\|}{\text{C}}}-R
\end{array}
$$

a Lipid (fat)

$$
\begin{array}{c}
\text{H}_2\text{C}-\text{O}-\overset{\displaystyle O}{\overset{\|}{\text{C}}}-R \\
| \quad\quad\overset{\displaystyle O}{} \\
\text{HC}-\text{O}-\overset{\displaystyle O}{\overset{\|}{\text{C}}}-R
\end{array}
$$

$$
\text{H}_2\text{C}-\text{O}-\overset{+}{\text{P}}-\text{O}-\overset{\text{H}}{\underset{\text{H}}{\text{C}}}-\overset{\text{H}}{\underset{\text{H}}{\text{C}}}-\underset{+}{\text{N}}(\text{CH}_3)_3
$$

a Phospholipid

Figure 10-1 Relationship between a lipid and a phospholipid

The true fats are the most abundant of all lipids; if liquid at ordinary temperatures, they are called "oils."

Fatty acids are a family of compounds consisting of chains of methylene (-DH2-) units, at one end of which is a carboxyl group (-COOH) which forms salts with sodium and with other earths and metals. Salts of fatty acids are called soaps. Fatty acids are transported to the various tissues through the bloodstream in complexes with the circulating albumin.

In mammalian tissues, most of the fatty acids vary in chain length from twelve to twenty-four carbons; but in fat the largest amount is made up of even-numbered, straight-chain compounds. Fatty acids also vary in their degree of unsaturation, which refers to the number of double bonds in a chain, points at which hydrogen has been removed so that adjacent carbons are linked together by two bonds instead of one. The saturated linkage between carbon is

represented thus: $-\overset{\text{H}}{\underset{\text{H}}{\text{C}}}-\overset{\text{H}}{\underset{\text{H}}{\text{C}}}-$

whereas double bonds are represented as: $-\overset{\text{H}}{\underset{}{\text{C}}}-\overset{\text{H}}{\underset{}{\text{C}}}-$

In higher animals, ingested triglycerides are large molecules broken down in the small intestine through the hydrolytic action of an enzyme present in the pancreatic secretion. Lipase activity has also been demonstrated in gastric juice. A partially purified preparation of pancreatic lipase causes the hydrolysis of the ester linkages of triglycerides in aqueous solution.

Triglycerides are split into glycerol and fatty acids by enzymes (lipases) and by alkali. Water must be present, three molecules adding to one triglyceride molecule. Lipases split the triglycerides at a slightly alkaline pH in a stepwise fashion. Diglycerides containing two fatty acids are first formed, part of these are then split to monoglycerides containing one fatty acid, and finally part of the monoglycerides are split to free glycerol and fatty acids. In the lumen of the intestine, absorption of soluble or emulsified mono-, di-, and triglyceride is so rapid that very little free glycerol is formed.

One view is that the absorption of the fatty acids follows the intestinal hydrolysis of fats, and this absorption is aided by the bile salts. In the intestinal wall the fatty acids are recombined with glycerol to form neutral fats. Another view is that bile salts are natural detergents and facilitate the entrance of the fats into the lymph vessels (lacteals). Estimates of the fat content of human lymph in the thoracic duct have been made, and they suggest that perhaps one-half of the ingested fat enters the lymphatic system; the rest is presumed to go directly to the liver via the portal blood.

(There is a laboratory exercise for the hydrolysis of lipids.)

Both the saturated fats and unsaturated oils take a much less active part in metabolic processes than the carbohydrates and proteins because of their insolubility in water. They usually accumulate in special fatty tissues under the skin, where they give the characteristic curves and shape to the body.

Demonstration: Identification of Fats

(1) The simplest way to identify fats is to make use of their ability to produce translucent marks on paper. With a medicine dropper add a drop of corn oil to the corner of a sheet of brown wrapping paper or any unglazed paper. To the opposite corner add a drop of water. When the fluids have evaporated, hold the paper up to the light and you will see a spot of grease still on the paper left by the corn oil, whereas there is absolutely no spot where the water was.

(2) To 3 ml of water in a test tube, add a drop of Sudan IV. Now add 1 ml of corn oil to the tube and shake thoroughly. You will find that the dye has been taken up by the oil.

(3) Instead of Sudan IV, you can add Sudan III and you will see the oil droplets turning pink as they take up the dye.

Demonstration: Hydrolysis of a Lipid to a Fatty Acid

(1) Label three test tubes 1, 2, and 3. (2) Into test tube #1, place 10 ml of water, a few drops of corn oil, 2 ml of soap solution (bile-salt solution can also be used), 5 ml pancreatin (pinch of pancreatin in 10 ml of water). (3) Into test tube #2, place 10 ml of water, a few drops of corn oil, and 2 ml of soap solution (or bile salts). (4) Into test tube #3, place 10 ml of water, a few drops of corn oil, and 5 ml of pancreatin solution. (5) Add a few drops of the phenolphthalein indicator into all three tubes so that the solution becomes light pink. Place all three tubes in a water bath. After some time tube number 1 will become colorless, showing that the corn oil has changed into fatty acid.

Since lipids must have a combination of an emulsifier and pancreatin to break down into fatty acids, what acts as an emulsifier in the human body when lipids are undergoing digestion?

AUDIO-VISUAL MATERIALS

At the moment there is nothing worthwhile available in this field, with the exception of a few isolated transparencies published by textbook publishers in conjunction with their books.

Nucleoproteins and Nucleic Acids

UNIT 11

At this point it is not necessary to say that proteins are most important biochemically; however, rivaling them for preeminence are giant molecules known as nucleic acids. There are two kinds of nucleic acids: deoxyribonucleic acid, usually called DNA, and ribonucleic acid, usually called RNA. DNA is always found in the nucleus of cells, while RNA may be found in the nucleus, but occurs mainly in the cytoplasm of the cell.

Very little of the DNA and RNA are present in the free form. They are combined with specific proteins. The DNA nucleoproteins thus formed usually contain 40-60 percent nucleic acid and the remainder is protein. The linkage of nucleic acid to protein is easily dissociated—the two components can easily be separated by passing an electric current through a solution of DNA nucleoprotein.

RNA nucleoproteins usually contain about 5-20 percent nucleic acid. The linkage of nucleic acid to protein is a more stable, covalent type. To separate the nucleic acid from the protein, one must either destroy the protein by denaturation or enzymatic digestion and recover the nucleic acid, or destroy the nucleic acid by digestion with the enzyme *ribonuclease* and recover the protein.

You have heard DNA called the *thread of life* since it is the essential constituent of the gene. Everything characteristic of a living organism, whether it be its size, its shape, the substances it can make, or its development from birth to death, is recorded in its DNA. No cell would know what to do next if its own peculiar DNA no longer existed. In short, DNA is the master planner of all life.

While DNA is the master planner, the position of RNA is equivalent to the production supervisor. The RNA is the blueprint or a working copy of the master plan DNA molecule, and is released after production by DNA in the nucleus to the cytoplasm. From its position in the cytoplasm, RNA directs the synthesis of a specific protein assembling the amino acids in a particular sequence in the polypeptide chain, which is dictated by the sequence of bases in RNA. Thus a single DNA molecule can duplicate itself and make a specific RNA molecule, which in turn can make a specific protein which is unique and different from other proteins in structure and function.

(There is little in the way of demonstrations that can be used for this unit, except for transparencies and an occasional model which should be used to help simplify the concepts in this unit.)

DNA

The class should know that DNA is only one of many nucleic acids, but it is the best known. DNA consists of many molecules of organic and inorganic substances, forming a giant molecule having a molecular weight of over one million. Such giant molecules consisting of repeated units are referred to as polymers by the biochemists. Each unit of DNA is composed of three different molecules: (1) The nitrogenous base, which is an organic nitrogen containing compounds with distinctive basic (not acid) leanings. The nitrogen base may be one of the purines, adenine or guanine, or one of the pyrimidines, cytosine or thymine. (2) Deoxyribose, which is a sugar unlike glucose or fructose, with a five-carbon skeleton but lacking one oxygen. (3) Phosphate, which is like any found in tissues or fertilizer.

The three substances are linked together in this order and the resulting group of molecules is known as a nucleotide. At each level of DNA, you will find a pair of these three molecules bound together with the nitrogenous bases forming the center, the sugar (deoxyribose) molecules attached to their outer edges, and finally the phosphates attached to the sugar. This group of molecules is

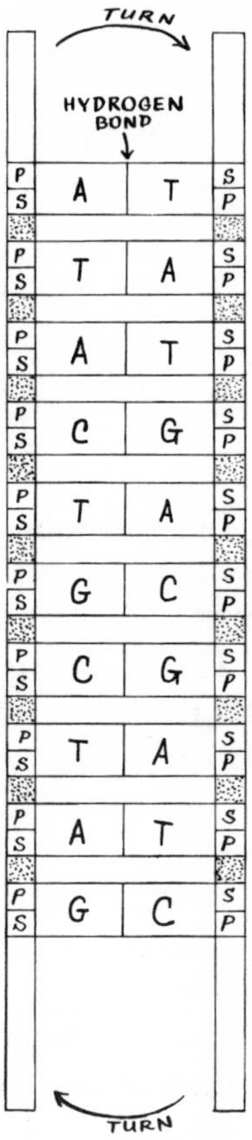

Figure 11-1 Schematic representation of a possible form of DNA as it would appear completely unwound. Key is as follows:

A - Adenine
T - Thymine
C - Cytosine
G - Guanine
S - Sugar (deoxyribose)
P - Monophosphate

Line between a pair of bases—weak hydrogen bond.
Stippled squares—strong chemical bond between one nucleotide and the one below or above it.

now called a nucleotide, and the nucleotide is the repeating unit in DNA (Figure 11-1).

DNA may consist of a single strand of nucleotides, but usually there are two stacks so arranged that their nitrogenous bases face each other, held in position by hydrogen bonds.

The four possible bases just mentioned, adenine or guanine (purines) or cytosine or thymine (pyrimidines), form the core of DNA, but they have a very definite arrangement at all times. Even when all required atoms are present, hydrogen bonding may not occur unless these atoms are properly disposed to one another (Figure 11-2).

If you examine the structure of the four nitrogenous bases which are found in the nucleotides of DNA, it becomes evident

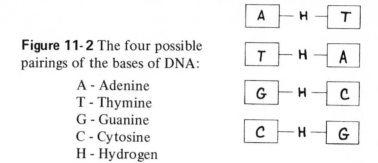

Figure 11-2 The four possible pairings of the bases of DNA:

A - Adenine
T - Thymine
G - Guanine
C - Cytosine
H - Hydrogen

that only certain ones can be paired to form hydrogen bonds. Thus it is not surprising that only certain nucleotides face each other along the two DNA strands.

The pair-forming nitrogen bases—the phenomenon is called "base pairing"—are adenine-thymine and guanine-cytosine. The positions within one of these pairs may be reversed, but the base pairs are always found together. The pairs are held together by hydrogen bonds. The hydrogen bond is a very weak one but plays an important part in the manner by which DNA is able to duplicate itself in the process of mitosis. It is the key to replication.

One interesting fact of this pairing is that the total amount of adenine is about equal to the total amount of thymine and, at the same time, the total amount of guanine is nearly equal to the total amount of cytosine.

The sequence of the paired bases sets up the code by which DNA can regulate the inherited characteristics and consequently the activities of the cell. It would seem that there are a very limited number of combinations of the base pairs, and that they could not possibly account for all of the inherited characteristics of a cell. However, if you stop to think of all the variations that are possible in stacking the base pairs, you will immediately realize that there are many hundreds of combinations.

During cell growth and before mitosis, DNA is rather evenly distributed in the nuclear material called chromatin. Before cell division this material is organized into threadlike chromosomes which split and provide the daughter cell with a set of chromosomes identical with those of the mother cell.

From what is now known of DNA we can imagine how this is done. Each chromosome has long spiral molecules of DNA. The DNA molecule "unzips" or splits like a ladder cut through the middle of its rungs. This leaves two threads, like a half-ladder with one upright and half-rungs. The half-rungs, in the case of the divided DNA molecule, are single bases, either A, T, G, or C, instead of pairs as in the whole molecule.

Thus the divided DNA molecule is rebuilt into the whole form, and an exact copy of the genetic code is passed on to the daughter cells. Through the sperm cells, the hereditary information is transmitted from generation to generation.

RNA

It is known that DNA produces ribonucleic acid (RNA) which, in turn, moves through the nuclear membrane out into the cytoplasm. RNA is similar to the DNA molecule which produces it, with three exceptions. First, it is usually a single strand instead of a double strand. Second, the base, thymine, is replaced by uracil; and third, the sugar, deoxyribose, is replaced by ribose.

One chain of DNA serves as a template to form a molecule of RNA. Each code letter determines its complementary code letter, establishing the order in the code letter in the RNA molecule; the U in RNA is the substitute for the T in the DNA as a bond of A. (DNA = A T G C and RNA = U A G C).

It is now believed that nuclear DNA produces three varieties of RNA, each different from the other in molecular size and function. One of these is ribosomal RNA, produced in the nucleus, but which moves to the endoplasmic reticulum where, together with protein, it forms the ribosomes. Each of these ribosomes is a protein-assembling machine.

A second form of RNA is a small molecule produced in some twenty varieties, one for each amino acid. These small RNA molecules are called *transfer RNA*. They leave the nucleus and relocate through the cytoplasm. Each variation of transfer RNA picks up a specific amino acid molecule. The transfer RNA molecules pick up amino acid molecules from solution in the cytoplasm, which they then transfer to the protein-assembling machine or ribosome.

The third type of RNA is messenger RNA, which carries

messages to the ribosomes about the coded information stored in the DNA molecule. Each message contains the instructions which the ribosome needs for the construction of a specific protein molecule. The messenger RNA molecules attach to the surface of the ribosome and thus provide a template for the assembly of amino acids into a protein molecule (Figure 11-3).

Amino acids in the cytoplasm are acted upon by enzymes and the high-energy molecules of ATP. The results of this action are that the amino acids are activated and prepared to attach themselves to smaller units of transfer RNA.

The transfer RNA consists of fewer molecules than messenger RNA and therefore has a smaller template. Each form of transfer RNA selects its particular amino acid and carries it to the template of the messenger RNA attached to the ribosome, depositing the amino acid in the proper site on the template. There are at least as many kinds of transfer RNA as there are amino acids. After depositing their amino acids on the ribosomal template, the units of transfer RNA go back to the cytoplasm to pick up again their specific activated amino acid. Meanwhile peptide bonds unite the main units on the template of the ribosome and a specific protein is formed, which is then released into the cytoplasm. Although it is not exactly known how transfer RNA moves to the template of the ribosome, depositing their amino acids in exactly the right sequence so that a protein will be formed, there is considerable evidence to show that the above description is correct.

NUCLEOTIDES AND NUCLEOSIDES

RNA and DNA can be hydrolyzed with acids or enzymes to yield basic building units. If the breakdown is stopped before it is complete, nucleotides and nucleosides are obtained.

Nucleotides and nucleosides are named after the base which they contain. The prefix "deoxy" is added if they contain deoxyribose instead of ribose. From previous mention in this chapter, you know that the nucleotide is composed of a nitrogenous base, a ribose of deoxyribose sugar, and a phosphate unit.

Each nucleotide is a larger molecule than the basic units of carbohydrates, fats, and proteins. Its structure is also much more involved than sugar, fatty acids, and amino acids.

Each nucleotide forms through two addition-elimination reactions. In the dual reaction a nitrogen base and a sugar unit join

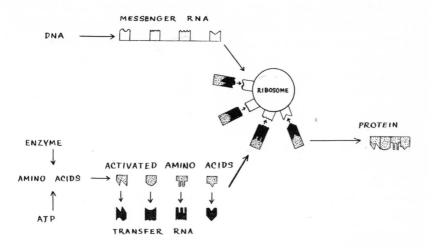

Figure 11-3 Diagrammatic representation of protein synthesis at the ribosomal site.

by elimination of a molecule of water. The phosphate group also joins to the sugar by elimination of water. By substituting the sugar, you can easily see how many nucleotides are possible.

Nucleotides are the basic structure of several compounds that are essential to life. Some vitamins, enzymes, and certain energy-rich compounds are modified nucleotides (Figure 11-4).

$$\text{Nucleic acid} \xrightarrow[\text{enzyme}]{\text{acid or}} \text{Nucleotides (base-sugar-H}_3\text{PO}_4)(-\overset{\text{B}}{\underset{|}{\text{P}}}\text{ -S- }\overset{|}{\underset{|}{\text{P}}}\text{-)}$$

$$\xrightarrow[\text{enzyme}]{\text{acid or}} \text{Nucleosides (base-sugar)(-S }\overset{\text{B}}{\underset{|}{|}}\text{ P)} \quad +\text{H}_3\text{PO}_4$$

Figure 11-4 Nucleotides

Demonstration: Extraction of DNA

At least two days before the demonstration, inoculate tubes of nutrient broth with a culture of E. coli and incubate.

On the day of demonstration transfer the nutrient broth containing the bacteria to centrifuge tubes. Balance the tubes and spin them for about three to four minutes at 2,500 rpm.

When you remove the tubes from the centrifuge, you will notice that there are two layers in the tube. The bacteria will be found in the lower fluid layer. Remove the upper fluid layer with a dropping pipette or medicine dropper and dispose of the material by placing it in a container of strong disinfectant.

Add an equal volume of sodium deoxycholate detergent solution to the bacteria suspension remaining in the centrifuge tube. You may use sodium lauryl sulfate (duponol) or any nonfoaming detergent instead of sodium deoxycholate.

With a square stirring rod briskly stir the solution for about ten minutes, shaking the tube at the same time. About three minutes from the start of the stirring add two drops of acridine orange and spin.

After ten minutes, gradually reduce the speed of the stirring and add absolute ethyl alcohol drop by drop until strings of DNA begin to form in the tube. The DNA can be removed with the square stirring rod.

Demonstration: Identification of Nucleic Acids

Peel off a few millimeters of the epidermis from an onion bulb. Place this transparent sample on a microscope slide and add a drop of 1N HCl. Heat the slide gently for a few minutes, adding hydrochloric acid as necessary, but do not boil the specimen or allow it to dry out. After heating, use a scrap of paper toweling to remove excess acid.

Place a drop of toluidine blue on the specimen and heat gently for a minute. Remove excess dye with a piece of paper toweling and then add a drop of water. Carefully place a cover slip over the specimen and place on a microprojector for the class to see.

The nucleic acids will be stained by the toluidine blue and everyone will note that the nucleic acids are only found in the nucleus of the cell.

Demonstration: DNA in Yeast

(1) From a good, growing culture of yeast cells, extract some cells and wash with distilled water. Centrifuge the suspension and

discard the supernatant. (2) Add 40% ethyl alcohol to the remaining cells. After two minutes, centrifuge the suspension again for ten minutes. (3) Rinse the cells three times with distilled water and centrifuge each time for five minutes. (4) Pour off the liquid after the third centrifuging and add 0.1M potassium hydroxide solution. Allow to stand for one hour at room temperature and then centrifuge again, this time for ten minutes. (5) Wash the cells several times with distilled water and centrifuge again for ten minutes. Discard the liquid portion. (6) Spread the yeast cells on a microscope in a thin film. Air dry and then stain for about two minutes with 0.1% ethyl alcohol and blot the slide dry. (7) Use a microprojector to show that there are blue masses within the yeast cells and then use an oil immersion microscope. (*Note:* Since there is so much time spent on centrifuging and other procedures, it might be best to use a double laboratory period and do some other work or lecture during the periods of waiting.)

Demonstration: DNA in Onion Tip Cells

(1) About one week before the demonstration place several onions in water, and when ready to use the onion root tips, cut off about 1 cm of the root tips and place in "fixing solution." (2) After they have been fixed, place them in a small vial of water and place a cheesecloth over the top of the vial fastened with a piece of string or rubber band. This will prevent the loss of the top when you place the vial under running water and let the water wash the tips for about five minutes. (3) Do not remove the cover but pour off the water and add hydrochloric acid to the vial, enough to cover the roots. (4) After ten minutes pour off the hydrochloric acid and wash under running water for a few minutes. (5) Pour off the water and cover the root tips with the dye leucofuchsin (Schiff's reagent). (6) After twenty minutes, wash off the excess dye by washing under running water for five minutes. (7) Add enough sodium bisulfite to cover the root tips, and have them remain in this solution for two minutes to bleach all parts of the cell containing DNA. (8) Remove root tips from vial, place one on a slide, and add a drop of mild acetic acid. Add cover slip and then, with a pencil or matchstick, roll the preparation to squash the root and separate cells from one another. (9) Use a microprojector to show that the DNA has been dyed purple.

AUDIO-VISUAL MATERIALS

FILMS

The Structure of Proteins and Nucleic Acids (H)
Structure and Replication (H)

SINGLE-CONCEPT FILM LOOP

Extracting DNA (A), (AA)

TRANSPARENCIES

A Coiled DNA Molecule (O)
An Uncoiled DNA Molecule (O)
Comparison of DNA and RNA (F)
Deoxyribonucleic Acid (F)
DNA (P)
DNA Set (A)
Interesting Facts About DNA (F)
Structure of DNA (F)

CHARTS

DNA Molecule (F), (P)
RNA (F), (P)

MODEL

DNA Molecule (F), (O), (CC)

KITS

DNA Molecule Construction Set (S)
Student DNA Model Kit (F), (S)

SOLO LEARNING

Introduction to DNA (A)

Energy and Life--Respiration

UNIT 12

The word "energy" is a common one that you have used often. What does it mean? The basic concept to have the class remember is that energy can move things, and all living things must have energy to survive.

Cells must spend energy to maintain and reproduce themselves. Even while you sleep, cells, tissues, and organs use energy. Some energy also is lost as heat through the skin. There is a minimum amount of energy necessary to maintain life. Since the movements of organisms require a large, constant, and carefully controlled supply of energy, the obtaining and use of energy through metabolism plays a large part in the process of living.

Although nonliving things also use energy, such as the movement of the sea and the movements of machines, this energy differs from the energy of the living world in that living organisms release and control their own energy.

(There are several demonstrations, such as the magnesium and the candle, that should be used at this point.)

Organisms get their energy from organic molecules through the life process called respiration. They extract it from chemical bonds in organic molecules. Green plants get the organic molecules they need by making their own, while animals get them in food, as you do yourself.

CELLULAR RESPIRATION IN ANIMALS

Basically, we use energy from the sunlight, but before it can be used the sunlight must be changed into chemical bond energy, which only green plants can do. All living things utilize energy from food. Plants store chemical energy in the food, which they alone among living things are able to synthesize. The energy in the plant is found in a glucose molecule and it in turn comes from the sun. Green plants in the process of photosynthesis change the radiant energy of sunlight into the chemical energy of glucose. This energy is later released in respiration.

As food moves through the organism, chemical reactions rearrange the molecules which make up the nutrients. These reactions break the chemical bonds which hold the molecules together. With the breaking of a bond, energy is released. The organism uses this energy to carry on the basic life processes.

The major fuels used by organisms are carbohydrates and fats, with little energy coming from protein. Carbohydrates are converted to glucose and the starting point for the release of energy begins here.

Burning in a fire and internal respiration in a living being are similar processes. Both are forms of oxidation. You can demonstrate the similarity between the two by blowing into a jar of limewater. The carbon dioxide you breathe out will turn the limewater milky. A burning candle also gives off carbon dioxide which will turn limewater milky.

If you heat a lump of sugar under the right conditions, it will catch fire and burn as a candle does. In cells the sugar also burns, but at a much lower temperature and much more slowly. Most important, its energy is released in a useful form and not wasted as heat.

The first stages in the breakdown of the sugar take place outside of the mitochondria. Midway through the respiratory cycle, the fuel molecules enter the mitochondria and it is this enzymatic activity that can be demonstrated or done as a laboratory exercise. Mitochondria are known as the powerhouses of the cell. Within these structures energy from food molecules provides the means by which cell processes are activated and sustained.

In the broadest sense, respiration includes all the processes involved in the release of this energy from food molecules.

It is necessary that the chemical energy in food be released in a slow, step-by-step process, not in one fast, wasteful surge as might happen if an organic molecule were burned in the laboratory. In this latter case, most of the chemical energy would be transformed to heat, which is quickly lost to the environment. In contrast, the energy released by respiration in the cell is converted step-by-step, releasing chemical energy which is available to the cell as needed.

The small packets of energy produced by the oxidation of glucose sugar are immediately used by the cell to change ADP plus phosphate to ATP. These ATP molecules then become a sort of currency that can be used as needed. Their chemical energy can be transferred to other processes that require energy. Thus these molecules serve as a storehouse of readily available energy.

Through respiration the cell is able to capture about 70 percent of the chemical energy within the glucose molecule. Glucose is one of the most common forms of carbohydrates, resulting from the action of chlorophyll. Since all the glucose produced by the chloroplasts cannot be immediately used by the plant cell, it is generally stored either as starch or cellulose. The two substances are in many ways similar; however, the bonds which hold together the glucose molecules are quite different in cellulose as compared with those of starch.

It is this difference in the bonding of the glucose molecules in cellulose and starch that explains an important difference among animals. The digestive systems of some animals contain enzymes that are able to break the bonds in cellulose, therefore reducing it to glucose from which they derive energy. These animals are herbivorous, such as cows and deer. Other animals lack these enzymes and cannot obtain energy from cellulose. This type of animal is carnivorous, and the human belongs to this class, as do meat-eating animals such as tigers.

Most animals have enzymes which break the bonds of glucose units in the starch as it is stored in plant cells. Carnivores obtain their starch from animal tissues in the form of glycogen. Glycogen is believed to be composed of some thirty molecules of glucose bound together in a starlike pattern. When glycogen is converted to glucose, the process is referred to as glycolysis.

If glucose is burned directly, it is converted to carbon dioxide and water. In addition to these two end products, a great deal of energy is released in the form of heat. The cell, although it needs heat for various chemical reactions, also needs other forms of energy, most of which are of a chemical nature. The cell accomplishes this by the slow breakdown or degradation of the glucose molecule in a series of steps. This matter of obtaining energy from glucose is involved and complex, but necessary since the cell must acquire its energy at a much slower rate of speed to meet physiological demands rather than to have the great burst of energy that would occur when plain heat converts glucose to carbon dioxide and water.

Two kinds of energy-releasing processes occur in cells. One is called *aerobic respiration,* and the other *anaerobic respiration* or *fermentation.* In aerobic respiration, oxygen is needed to bring about a combustion of materials (usually sugars) in the cell to release energy. In anaerobic respiration, energy is released in the absence of oxygen. The essential chemical event in anaerobic respiration is the rearrangement of sugar molecules. The energy obtained from both kinds of respiration is used to make adenosine triphosphate (ATP). Also, carbon dioxide and heat energy are released.

Aerobic respiration is a more complex chemical process than anaerobic respiration. Many steps are involved. The anaerobic respiration of one molecule of glucose yields two molecules of ATP. The aerobic respiration of one molecule of glucose yields approximately thirty-eight molecules of ATP. The aerobic respiration of glucose can be summed up in the following empirical equation, which is a summary and by no means expresses the actual chemical events of aerobic respiration.

$$C_6H_{12}O_6 + 6O_2 \xrightarrow{\text{enzymes}} 6CO_2 + 6H_2O + \text{approximately } 38\sim P$$

AEROBIC RESPIRATION

High energy biochemical compounds are very often rich in phosphorus. One of the most important of these substances is adenosine triphosphate (ATP). ATP consists of a molecule of adenine plus ribose plus three atoms of phosphate. It is the third molecule containing phosphate which gives ATP its high energy capacity. ATP is generally formed from adenosine diphosphate, which has one less phosphate attached to it.

Adenosine   Diphosphate   (ADP) + P→Adenosine   Triphosphate (ATP)

There is experimental evidence that absorption of sugar involves the combining of glucose molecules with phosphate ions. This reaction, called phosphorylation, requires special enzymes and energy. After the glucose phosphate complex enters the cell on the lumen side, some of the molecules are used by the cell for their metabolic processes. Others are secreted opposite their point of entry into the bloodstream. Here the glucose-phosphate combination is split, i.e., dephosphorylation. The free glucose passes out of the cell, whereas the phosphate ions remain and may be used to form glucose-phosphate complexes.

Compounds that uncouple phosphorylation, i.e., prevent union of sugars and phosphate, might interfere with active transport processes.

There are two basic ways adenosine triphosphate is synthesized. One is a method by which the phosphate group is transferred to the ADP molecule directly from a degradation product of glucose. Another method which accounts for the majority of the synthesis of ATP is referred to as the oxidative respiratory chain.

One basic step which must occur in the respiratory chain before the synthesis of ATP is that nicotinamide adenine dinucleotide (a coenzyme) has to accept or receive two hydrogen atoms from a substrate. This coenzyme is often called NAD or DPN. In the presence of NAD, malic acid transfers two of its hydrogens attached to the first carbon atom, with the result that $NADH_2$ and oxalo-acetic acid are formed. The NAD, with its hydrogen, is now prepared to enter a series of reactions known as the respiratory chain, involving the degradation of glucose. The released hydrogen combines with oxygen to form water as oxygen serves as a hydrogen acceptor. The remnants of the original glucose molecule are reorganized to form carbon dioxide.

The overall summary equation for aerobic respiration of glucose is:

$$C_6H_{12}O_6 + 6O_2 \xrightarrow{\text{enzymes}} 6CO_2 + 6H_2O + \text{energy (38 ATPs)}.$$

This formula is the same as for photosynthesis except it is in reverse. You should remember three aspects of this which are important. First, some of the hydrogens involved in the formation of the water are transferred from $NADH_2$. Second, the energy represented is derived from ATP. Finally, a great deal of oxygen is required for this conversion of glucose to its end products and release of energy. Therefore, this portion of glucose breakdown is referred to as an aerobic reaction. The aerobic portion of glucose degradation follows an anaerobic phase in which no oxygen is required. This type of chemical breakdown of glucose is glycosis.

During the anaerobic phase, which does not involve molecular oxygen, glucose is converted to two molecules of pyruvic acid, and the released energy is used to synthesize four ATP molecules. However, since the energy of two ATPs is necessary to activate the reactions, there is a net gain of only two ATP molecules during the anaerobic phase.

The reactions of the aerobic phase continue the process of energy release by breaking down the pyruvic acid, eventually forming carbon dioxide. During this process thirty-six ATP molecules are synthesized. The net gain from the aerobic respiration of glucose is therefore twenty-seven molecules of ATP.

ANAEROBIC RESPIRATION

You have been talking about anaerobic respiration to the class, but now it will be discussed in greater detail. Almost all animals carry on aerobic respiration. However, certain animal cells can carry on respiration anaerobically when oxygen is deficient or absent.

When it occurs in animals, anaerobic respiration leads to the formation of lactic acid and a relatively small amount of ATP. Lactic acid production may be summarized as follows:

$$\text{glucose} \xrightarrow{\text{enzymes}} \text{2 lactic acid} + \text{2 ATP}$$

Lactic acid still contains much unreleased potential energy and the net gain of ATP is two molecules. This is a considerably less efficient energy-releasing system than the aerobic respiration of glucose.

Vigorous activity of human voluntary muscles may produce an oxygen deficiency, which leads to a certain amount of anaerobic respiration and the accumulation of lactic acid in the cells. Lactic acid build-up is associated with fatigue.

(There are several demonstration and laboratory exercises that can be done on the topic of fermentation and anaerobic respiration.)

ENERGY FROM FATS AND PROTEINS

Your energy comes not only from carbohydrates but also from fats and proteins. The fat molecule, a neutral fat, through the action of hydrolytic enzymes is broken up into four units, three of which are fatty acid molecules and one of glycerol. The reaction is:

$$\text{Neutral fat} \xrightarrow[\text{enzymes}]{\text{hydrolytic}} \text{3 fatty acids + 1 glycerol}$$

Every neutral fat molecule is broken up into three fatty acids. Each fatty acid in turn is broken up into ten or more two-carbon fragments, each one of which may produce eleven molecules of ATP. In addition, the one molecule of glycerol produces even more ATPs.

Proteins also produce energy, but percentagewise they are not as important as carbohydrates and fats. When enzyme systems break down proteins, most of the resulting amino acids are used to synthesize other proteins. Actually very few of them are used for energy formation.

We have been talking about ATP molecules being produced, but where does all this take place? For the most part they are synthesized in the mitochondria, where sugar products are burned and their energy used to form ATP.

Neutral fats are degraded in the hyaloplasm, which is the cytoplasm of a cell outside the endoplasmic reticulum. The neutral fats are degraded to their constituent part, fatty acids and glycerol, which in turn pass across the mitochondrial membranes and undergo subsequent reactions until they are broken down into acetyl coenzyme units and enter the Krebs cycle. The same is true of proteins. They are broken down into amino acids in the hyaloplasm and then passed into the mitochondria to undergo the final steps.

It is known that the mitochondria are involved with the energy formation in the cell since there is a direct relationship between the number of them and the activity of the cell. The more active the cell, the greater the number of mitochondria. It also has been found that mitochondria group together in the area of the cell where the greatest amount of biochemistry is taking place. Finally, it has been observed that during periods of rest the mitochondria actually increase in size, as they accumulate more and more ATP; after long periods of activity, they definitely become smaller.

CELLULAR RESPIRATION IN PLANTS

Aerobic respiration occurs in the presence of free molecular oxygen and will be fully discussed in the unit on photosynthesis.

Anaerobic respiration does not require free oxygen. Since molecular oxygen is not used in the release of energy from organic compounds in anaerobic respiration, it takes place whether or not oxygen is present. However, in the continued absence of oxygen, ethyl alcohol or lactic acid may be produced by a process called fermentation.

Alcoholic Fermentation   This process begins with the formation of pyruvic acid by the same sequence of enzyme-catalyzed steps that produced it in animal tissues. If molecular oxygen is lacking, some simple plants, such as certain bacteria and yeasts, carry on alcoholic fermentation, producing ethyl alcohol and carbon dioxide as products of the reaction. The organisms derive their energy from the glucose bonds. A simplified summary equation for the alcoholic fermentation of glucose is:

glucose → pyruvic acid → ethyl alcohol + carbon dioxide + energy

Most of the released energy of glucose still remains in the bonds of alcohol. Alcohol fermentation has value in such industries as baking, brewing, and wine making.

Lactic Acid Fermentation   Lactic acid bacteria and certain molds can carry on lactic acid fermentation. In this process the bacteria or molds also obtain their energy from the bonds in glucose and lactic acid in the product of reaction, as seen earlier in the chapter. Lactic acid fermentation has commercial value in the food, fermentation, pharmaceutical, and chemical industries.

Since most of the C-C and the C-H bonds of the compound remain intact at the end of both alcoholic and lactic acid fermentation, only a relatively small amount of energy is released in comparison to the amount released in aerobic respiration. However, anaerobic respiration provides some energy to support lower forms of life, such as bacteria and yeasts, whereas aerobic respiration provides the energy needed by higher plants.

Demonstration: Products of Oxidation

Hold a clean glass plate directly above a burning candle. After a few seconds examine the surface of the plate and you will find a black deposit on the glass, a result of oxidation. If you repeat the experiment after placing the candle in a bottle of pure oxygen, the flame will die out and you will find that the plate does not have a layer of black deposit on it.

Demonstration: Another Oxidation Demonstration

Shape some butter into a small mound in a watch glass or small dish. Place a cotton wick in the butter and ignite. Hold a small cold mirror over the butter candle and look for droplets of water on the surface. Now dry the mirror and cool it. Breathe on the mirror and again examine its surface for condensed water. You will note that there is a connection between these two experiments.

Demonstration: Action of Sulfuric Acid on Sugar

Place a lump of sugar on a watch glass. Allow a few drops of concentrated sulfuric acid to drip on it. (Make certain that the sulfuric acid does not touch the skin.) At first the sugar turns brown, then it becomes black.

Sugar is an organic compound that falls into the category of carbohydrates. Carbohydrates contain carbon, hydrogen, and oxygen. Sulfuric acid has an extremely strong affinity for water. When in contact with a carbohydrate the acid extracts the water, leaving the carbon behind. It is this carbon which makes the sugar appear black.

Demonstration: Combustion of Magnesium

Cut an 8 cm length of magnesium ribbon. Holding one end of the ribbon with a pair of crucible tongs, insert the other end into the flame of a Bunsen burner and hold it there until it begins to burn. Simultaneously, catch some of the "smoke" that rises from the flame in an inverted porcelain cup held above it. (*Make certain that the pupils do not look at the flame for too long a period of time.*) The magnesium ribbon will burn with a dazzling white light. The residue is brittle and powdery because this is a rapid oxidation reaction where a temperature of about 2,500° C is attained. The extremely bright flame is a result of the extremely high temperature.

Demonstration: Fermentation, Carbon Dioxide Production

(1) Make a yeast suspension either by crumbling half a yeast cake into a beaker containing 100 ml of distilled water or by adding dried yeast cells to the water. Divide the suspension into two 50 ml portions and to one portion add five grams of glucose or some molasses. Mix the solution thoroughly. (2) Fill three fermentation tubes with the glucose-yeast mixture until the arm and bulb are filled. Use the yeast suspension alone to fill a fourth fermentation tube. (3) Label #1 the fermentation tube with the yeast only. Add five drops of brom thymol blue to one of the fermentation tubes containing the yeast-glucose suspension and label it #2. (4) Plug all four tubes with absorbent cotton and set them into an incubator at 30° C. Examine them in about an hour and again the following day. You will find that tube #1 had no odor or gas produced. Tube #2 had a fruity alcoholic odor and gas. The blue color has changed to yellow. In tube #3, the column of air has almost completely disappeared after you added about 2 ml of 10% solution of potassium hydroxide, since the potassium hydroxide is carbon dioxide absorbent. In tube #4, the column of air remains since nothing was done to it.

You know that carbon dioxide is produced because of the change of pH, and also by the action of the potassium hydroxide.

Laboratory Exercise: Yeast Anaerobic Respiration

Aim: To study the respiration of yeast, especially anaerobic respiration.

Methylene blue is used to measure the changes that occur as the yeast culture grows. Glucose will be the source of energy for the yeast. Methylene blue is blue when in an oxidized state and colorless when reduced. If this dye changes from blue to colorless, it is indicative of the fact that the dye is being reduced and some other chemical is being oxidized. Yeast is a typical fermenting organism and therefore ideal for the study of anaerobic respiration, also known as fermentation. Anaerobic respiration releases energy for use by this organism. This process makes life possible in an environment that lacks free oxygen.

Materials Needed: Yeast; 5% glucose solution; seven test tubes; test tube rack; water bath, both cold and warm; methylene blue (0.05% of which is made by adding 0.5 gm methylene blue to 1000 ml of distilled water); 10 ml graduated cylinder; medicine dropper.

Procedure: (1) Make a suspension of yeast by adding a package of dry yeast to 100 ml of 5% glucose solution. (2) Label seven test tubes from 1 to 7 and add 5 ml glucose solution to each. (3) To tubes 1, 3, and 5, add 1 ml of yeast suspension, shaking the suspension thoroughly before withdrawing a sample. (4) To tubes 2, 4, and 6, add 0.5 ml of yeast suspension and 0.5 ml of 5% glucose. (5) Tube seven will serve as a control with no yeast in it. (6) Shake the tubes thoroughly to mix the contents. Keep tubes 1, 2, and 7 at room temperature. (7) Place tubes 3 and 4 in a cool water bath of water and ice (10-15° C). (8) Place tubes 5 and 6 in a warm water bath (37° C). (9) Wait about five minutes and then add five drops of 0.05% methylene blue to each of the seven tubes. All tubes will be blue, but #7 will be clearer. Note the time required for the disappearance of the blue color. A slight blue ring will remain at the top of the liquid.

Observations and Conclusions: The amount of time involved in the disappearance of color is an index of density of population; the faster the disappearance, the faster the yeast will grow. The yeast grows fastest in tube #5. The order of time for color disappearance follows.

Order of color change	Number of tube
first	5
second	6
third	1

Order of color change	Number of tube
fourth	2
fifth	3
sixth	4
no reaction	7

Warm conditions caused the yeast to grow best.

Laboratory Exercise: Mitochondria

Aim: To study the work of mitochondria.

Mitochondria are known as the "powerhouse" of the cell. Within these structures energy from food molecules provides means by which cell processes are activated and sustained. Mitochondria are the smallest structures that can be seen with the light microscrope. Special staining techniques are needed, such as Janus green B, which acts as an acceptor of hydrogen from food molecules that are being degraded. In its oxidized state Janus B is green, while it is colorless in its reduced state.

Materials Needed: Stalk of celery; microscope slide and cover slip; compound microscope; 5% sucrose solution; 0.001% Janus green B; forceps; razor blade; paper toweling.

Procedure: (1) With the use of a razor blade cut a slice 1 cm thick from a celery stalk in a transverse slice. Put a drop of sugar solution on a slide and place the celery slice, cut side downward, into the solution. (2) Locate the two strings that run through the celery slice. Hold one end of the slice with the forceps and cut through the strings. This will leave a transparent cube. Tease the thin inner membrane away from this cube. Remove the celery pieces you do not need from the slide and add another two drops of sugar solution to the membrane. (3) Cover with a cover slip and, using the high and low power of the microscope, study the mitochondria. (4) Withdraw the sugar solution from under the cover slip with a square of paper toweling. At the same time run some Janus green B under the opposite side of the cover slip. Study the activity of the mitochondria under the microscope. (5) The mitochondria will be blue-green, small rod-shaped or spherical structures. As you watch the material on the slide, you will see that the mitochondria lose their color and change to gray or colorless. (6) The decolorization of the mitochondria did not take place immediately, since the first stages in the breakdown of the sugar take place outside of the mitochondria. Midway through the respiratory cycle the fuel molecules enter the mitochondria, and it is this enzymatic activity that was demonstrated.

AUDIO-VISUAL MATERIALS

FILM

Nature of Life; Energy and Living Things (K)

FILMSTRIP

Respiration (The Living Cell Set) (W)

SINGLE-CONCEPT FILM LOOPS

Aerobic Respiration (A)
Effect of Temperature (Respiration) (A)
Fermentation (Respiration) (A)
Measuring Rate of Respiration in Mitochondrion (O)
Role of ATP in Muscle Contraction (A)

TRANSPARENCIES

ADP and ATP (F)
The ATP-ADP Cycle (O)
The Krebs Cycle (F)

MODEL

Mitochondrion (A)

KITS

Metabolism Experiment Kit (A), (O), (S), (CC), (EE), (FF)
Respiration (A)

REPRINT

Energy Factory (X)

SOLO LEARNING

Introduction to Biological Oxidation and ATP (A)
Introduction to Energy Cycle and Trophic Levels (A)

Animal Digestion

UNIT 13

You might think that your students all know the definition of the word *food,* but you would be surprised at the
misinformation that people really have about the word. Foods
differ from other materials taken into the body.

Foods or nutrients must be able to do three things; in the
first place, they must be able to supply energy to the body by
acting as a fuel. The energy is acquired through cell respiration.
Second, they can be used to increase the amount of protoplasm
and act as structural materials for growth, maintenance, and the
repair of the organism. Third, they can be stored and used as a
reserve supply on which the body can draw in times of emergency.
They help in the operation of the various metabolic processes.
Thus a food may serve as a source of immediate energy or may be
temporarily stored for future energy and repair requirements.

There are six groups of nutrients: water, mineral salts,
carbohydrates, fats, proteins, and vitamins. No one group of food
materials satisfies all the requirements of a perfect food, because
each lacks something essential for complete and proper nutrition.

Water is necessary for all living things; without it, plants
could not manufacture their food and animals would be unable to
exist. Water is the only fluid in which the materials present in the
human body can dissolve. Without enough water the cells would
lack food and oxygen, and wastes would collect in them. The
water and materials dissolved in them maintain the osmotic
pressure of the cells.

Water acts as the carrier of materials between the cells and their surroundings. Water also acts as a temperature control through evaporation from the surfaces.

(You can test various foods' water content by doing the demonstrations at the end of the unit.)

Mineral Salts   Minerals or inorganic salts function in a variety of ways as structural materials and as metabolic regulators. They play an important role in the regulation of cellular functions; each exerts its own osmotic action and is able to react with or combine with other compounds to form new materials. Everybody eats sodium chloride when he salts his food. The sodium chloride determines the electrolyte content of body fluid and may appear as constituents of blood, cell proteins or sweat. Water balance and salt balance are closely interrelated. (Tests for minerals should be done here.)

Carbohydrates   Carbohydrates play several important roles in the body. In the first place, their oxidation quickly releases a large amount of energy. The liver and muscles are the principal places of storage. Glycogen alone may account for 20 percent of the dry weight of the liver. Glucose is converted in the mitochondria into glycogen to be stored, and in times of need the glycogen is reconverted into glucose, entering the bloodstream and carried where needed.

Carbohydrates may also be converted into fats and stored in that form. They also act as protein savers in that the carbohydrates are oxidized first rather than proteins to provide energy, since proteins serve other purposes.

Lipids   Lipids can be oxidized for energy, and also are used for energy storage and body building. Most of our digested fat is oxidized to produce energy since, pound for pound, fats produce more than twice as much energy as either carbohydrates or proteins.

Proteins   The principal uses of proteins are to repair and replace protoplasm and to form new living material. Nitrogen is the essential element in the protein molecule; without the nitrogen no protoplasm could exist. Energy is also supplied by proteins. However, there is no provision for protein storage, so proteins not immediately used in the formation of new protoplasm can be built into the protoplasm of existing cells or transformed into sugars and fats.

(You need not do any tests for carbohydrates, lipids, or proteins unless you have not done them previously.)

Vitamins   Vitamins are essential food substances although they are required in only minute amounts. However, without them the body cannot use some of the foods it takes in or carry on numerous vital functions necessary for the maintenance of normal conditions.

DIGESTION

If nutrients passed down the digestive tube without change, there would be no value in eating. Digestion is the process whereby complex food molecules are broken down by chemical action into simple molecules that can be absorbed by the body.

Digestion is partly mechanical as the food is broken down into small pieces by chewing, crushing, squeezing, and mixing. It is also partly chemical, as nutrients are changed by hydrolysis into simpler materials.

Very large molecules, such as starch, proteins, and fats, are insoluble in water. These molecules may appear to be soluble since they are able to form colloidal solutions. However, they still are unable to pass through the small holes of semipermeable membrane. Most of the absorption of the hydrolyzed or digested food is carried on in the intestines.

Digestion occurs as follows: Food molecules are too big to pass through the digestive tract wall and must be broken down. After being broken down the molecules are absorbed, and after absorption they are resynthesized into large molecules in the body. The digestive tract acts as a typical semipermeable membrane and prevents large molecules from coming into the circulating fluids of the body, and the large molecules of the fluids are prevented from leaving the body. Small molecules, such as glucose, are in the simplest form and can be assimilated without any digestive action.

To be more specific, the digestive system of the mammal is the result of the complicated selected process of evolution. Structurally it is adapted to fulfill the needs of the mammalian diet in breaking down food material to the level where it can be used as a direct energy source in cellular metabolism or incorporated as building material in the cell. In principle, however, the mammalian digestive system is directly analogous to the food vacuole of the amoeba. In fact, many of the active sites of the enzymes which are involved in the breakdown of foodstuffs are essentially the same in all organisms.

(At this point do the intracellular digestion in the paramecium.)

The general pattern of food breakdown is one of progressive degeneration. The degradation to final usable form of each nutrient type must be completed by the time the food reaches the middle section of the small intestine, where most of the absorption into the body takes place. Therefore all the digestive enzymes are found to be secreted by structures which precede this area.

Carbohydrate Digestion   Amylase is found in saliva and is responsible for the breakdown of plant starch (polymerized glucose) to the disaccharide, maltose, a dipolymer of glucose. Pancreatin is a conglomerate of digestive enzymes extracted from the pancreas. The pancreas is a gland that lies in the curve between the stomach and the duodenum. It acts as both an *exocrine* and an *endocrine* gland. As an endocrine gland it secretes insulin, a substance important in glucose metabolism. As an exocrine gland it secretes a host of digestive enzymes that enter the intestine through the pancreatic duct. Among these enzymes is pancreatic amylase.

Protein Digestion   Protein digestion begins in the stomach. First, *rennin* acts to change the consistency of milk, which contains a great deal of protein, but does not alter the bonds of the protein structure. *Pepsin*, which enters the stomach in the inactive pepsinogen form, is activated by the acidic condition of the stomach and reacts on several specific sites of the protein molecules, breaking them into large polypeptides.

The material empties into the intestine. The *trypsinogen* secreted by the pancreas is activated by the basic condition of the duodenum. In turn, *trypsin* activates *chymotrypsinogen* to its active form, *chymotrypsin*. Each of these digestive enzymes acts at a different site on the protein molecules, thus breaking them into smaller polypeptides and dipeptides. *Dipeptidases,* secreted by the intestine and pancreas, are responsible for the final breakdown of dipeptides into amino acids.

Fat Digestion   Because of the high degree of insolubility of fats in water, the body has devised indirect means of breaking down lipid material. Fat in a water environment will take a spherical form, yielding a maximum volume to a minimum of surface area. This is an adverse condition for chemical reactions because these reactions are dependent on the availability of a substrate on which to act. *Bile salts* act as emulsifying agents, reducing the surface tension of fat and causing the fat globule to break up into smaller particles, which remain in suspended form. By increasing the number of spherical forms with accompanying reduction in size, the surface area is increased.

Digestion in the Mouth   After the food has been chewed and crushed by the teeth, saliva is secreted by the six salivary glands, the paratoid, the submaxillary, and the sublingual. About 1,500 ml of saliva are secreted daily by an individual. Saliva is more than 99 percent water, but the small amount of solids in this liquid is important. Saliva is ordinarily neutral, with a pH of about 7.0. When stimulated by food the saliva becomes acidic, with a pH of 6.5 to 6.8. The optimum pH of ptyalin, an enzyme in saliva, is 6.6. Mucin is also found in the saliva. It is a glycoprotein composed of carbohydrate and protein, and makes the saliva slippery to aid in swallowing. The mineral salts in the saliva constitute the buffer systems of the mouth. There are carbonate buffers and phosphate buffers which help keep the pH of saliva around 6.6 and thus aid in the production of active ptyalin.

Ptyalin is the most important enzyme in the mouth. It is an amylase, acting on starch. The action of the ptyalin breaks down starch through a series of reactions to form maltose: (1) starch + water = amylodextrin + maltose, (2) amylodextrin + water = erythrodextrin + maltose, (3) erythrodextrin + water = achrodextrin + maltose, (4) achrodextrin + water = maltose + maltose.

(There are many demonstrations and laboratory exercises to illustrate digestion in the various parts of the digestive tract.)

Digestion in the Stomach   The stomach is a storage organ as well as a digestive organ since undigested food may remain there for several hours. The length of time the food remains in the stomach is in part a function of the type of food eaten. For example, fatty foods, in general, remain longer in the stomach than carbohydrates since they are, in general, harder to digest.

The inner lining of the stomach is a mucous membrane called the gastric mucosa. The gastric mucosa contains the long tubular gastric glands that secrete the gastric juices used in digestion. It has been estimated that some 35,000,000 of these glands line the normal stomach.

About two to three liters of gastric juice are secreted daily by a normal adult. There is a constant gastric juice flow, but when food enters the mouth or is smelled or seen, the nervous stimulus causes a marked increase in the gastric juice flow. It is believed that the stimulus causes the production of a hormone known as gastrin, which stimulates the production of gastric juice.

The gastric juice contains several substances, of which the proenzyme *pepsinogen* is the most important. The pepsinogen is converted by the hydrochloric acid into the active enzyme *pepsin*, which is a proteolytic enzyme capable of hydrolyzing proteins. For this reason, the chemical action on proteins is the most important digestive reaction occurring in the stomach.

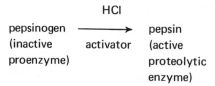

Pepsin does not digest protein into amino acids but changes proteins into proteoses and peptones.

Remind the class that the gastric juice is the most acidic fluid of the body. This is due to the presence of hydrochloric acid. Although HCl in concentrated form can kill, protection seems to be offered by the salts of sodium and potassium, especially when they are combined with chlorides and bicarbonates. The range of acidity of gastric juice is from a pH of 1.7 to 0.3.

The HCl functions in several different ways during the gastric digestion. In the first place, its presence is necessary for the formation of pepsin from the proenzyme, pepsinogen. Second, pepsin is unable to act in the digestion of proteins unless the gastric contents remain acidic. A third activity of the HCl is to destroy bacteria that might have entered with the food, and the fourth property of hydrochloric acid is that it dissolves some of the salts we eat.

As the gastric juice is secreted the bloodstream tends to become slightly more basic than normal. This is especially true after a meal.

Because a great deal of free strong acid is in the gastric juice, undoubtedly some hydrolysis of disaccharides, such as sucrose, maltose, and lactose, occurs. However, this chemical action depends to some extent on the length of time the food remains in the stomach. In the presence of hydrochloric acid, sucrose changes to glucose and fructose, maltose to glucose and glucose, while lactose changes to galactose and glucose.

Gastric lipase is another enzyme in the stomach. This has the ability to break fats down into fatty acids and glycerol. There appears to be a small amount of this enzyme in the stomach, but its action is very weak because it is destroyed when the acidity of the stomach becomes higher than 0.2 percent.

$$\text{any oil or fat} \xrightarrow[\substack{\text{gastric} \\ \text{lipase}}]{+3H_2O} \text{glycerol} + \text{fatty acid}$$

Rennin is a proteolytic enzyme found in the gastric juice secretions of certain young mammals. There is doubt that any rennin is to be found in an adult's gastric juice. Rennin has about six times the coagulating power of pepsin. However, the hydrochloric acid and pepsin in the stomach adequately cause the precipitating and hydrolysis of protein in milk. (If you wish you can use some rennin powder for illustration.)

As the digestive process continues in the stomach the contents become liquefied, and when properly liquefied the muscular contractions of the stomach wall force them toward the pyloric opening, and, by peristaltic waves, into the duodenum. The liquefied material is called chyme.

Digestion in the Small Intestine   As the food leaves the stomach it enters the small intestine, a narrow tube only one inch in diameter but over twenty feet in length. The inner surface of the small intestine is covered by mucous membranes, and the indentations along the inner wall are the site of millions of glands that make the intestinal juices. But certainly as important as the intestinal juices are the secretions that pour into the duodenum from the liver and pancreas.

The *liver* is the largest gland in the body. Internally, the liver is divided into lobules, and these in turn are made up of cords of liver cells that radiate out from a central vein in a most irregular manner. Each cord consists of two adjacent rows of cells, and within these cells take place the many chemical activities that make the liver an intricate mechanism concerned with digestion and metabolism. Our chief interest here is the function of the liver in secreting bile.

There appears to be a continuous flow of bile, about 500 to 1,000 ml daily, which varies according to the diet as protein foods cause a greater production of bile than carbohydrates. Although bile has little digestive power itself, it is a great aid in digestion.

The bile is made up of three substances: the bile pigments, the bile salts, and cholesterol. The cholesterol and bile pigments may be regarded as excretions, since they leave the body. The bile pigments are formed as a result of the breakdown into bilirubin in various organs and tissues such as the spleen, liver, connective tissue, and bone marrow.

The bile salts are sodium glycocholate and sodium taurocholate which are concerned with the digestive process. The digestive system uses these salts as a detergent to surround droplets of fats with a layer of "soap" so that the droplets do not fuse together. An emulsion is formed. The digestive enzymes can thus break down the oil. A second function of bile salts is to make water-soluble those fatty acids that do not readily dissolve in the intestinal fluid. A third function of these salts is to activate steapsin by converting the inactive prolipase into the active steapsin. Finally, the fat-soluble vitamins A, D, E, and K could not be properly absorbed in the body if fat particles were not emulsified properly.

As the chyme is meted out into the duodenum in small portions by means of the pyloric sphincter, it is quickly mixed with the flow of pancreatic bile and intestinal juices. These juices are alkaline and neutralize the acidity of the chyme. Thus the intestines are alkaline.

The pancreas is an elongated, somewhat triangular gland, which contains two types of glandular tissue. One type predominates and is the source of pancreatic juice. Isolated groups of cells, the Islands of Langerhans, produce insulin, a hormone which plays an important role in the carbohydrate metabolism.

Two substances are involved in the production and flow of pancreatic juice. One is a protein hormone found in the duodenal mucosa. This hormone is a pancreatic secretion which is influenced by the chyme and the free hydrochloric acid from the stomach. The pancreatic secretion is carried through the bloodstream to the pancreas, where it stimulates the flow of pancreatic juice. The second hormonal substance, called pancreozymin, is found in the duodenal tissues, and its function is to stimulate the production of the pancreatic juice. Thus pancreozymin causes the pancreas to produce the necessary enzymes, and pancreatic secretion causes the actual flow of these enzymes into the duodenum.

Activated pancreatic juice is one of the most powerful digestive juices in the body. The liquid contains powerful enzymes such as trypsin, chymotrypsin, carboxypeptidase, amylopsin (an amylase), and steapsin (a lipase). In addition, several enzymes in trace quantities are capable of hydrolyzing certain proteoses, peptones, and polypeptides. The activated pancreatic juice also contains a high proportion of sodium bicarbonate and disodium hydrogen phosphate which render the juice quite alkaline, within the 7.5 to 8.0 pH range.

Trypsin and chymotrypsin are secreted as the inactive proenzymes, trypsinogen and chymotrypsinogen, respectively. They are activated by the enzyme *enterokinase* in the intestinal mucosa. The enterokinase catalyzes the conversion of trypsinogen to trypsin, which in turn converts chymotrypsinogen into chymotrypsin.

Trypsin and chymotrypsin have an optimum of 8.0 pH. Both enzymes are similar in their activity in that they both hydrolyze proteins to the proteose-peptone stage, although they are not substitutes for each other and often one acts where the other left off.

An enyzme formerly known as erepsin has now been shown to contain many different enzymes, each one acting upon a particular substrate. The activity is concerned with the hydrolysis of proteoses, peptones, and polypeptides to a-amino acid. We know that erepsin is a mixture of peptidases, one of which is carboxypeptidase, which hydrolyzes peptides by removing the last amino acid group.

The chief carbohydrate-hydrolyzing constituent of pancreatic juice is the pancreatic amylase known as amylopsin, which has an optimum pH of 7.1; its activity is almost identical to that of ptyalin, except that amylopsin is so powerful that it will even hydrolyze raw or uncooked starch.

$$\text{starch} \xrightarrow[+\text{H}_2\text{O}]{\text{amylopsin}} \text{dextrins} \xrightarrow[+\text{H}_2\text{O}]{\text{amylopsin}} \text{maltose}$$

The important lipolytic enzyme present in the pancreatic juice is known as steapsin (pancreatic lipase). As the pancreatic juice enters the intestine, the inactive proenzyme *prolipase* is converted into the active steapsin. Bile salts can affect this conversion. Steapsin hydrolyzes the emulsified fat into fatty acids and glycerol.

INTESTINAL ENZYMES

Digestion of the chyme is completed by the intestinal enzymes. There appears to be two types of secretions. One secretion is independent of the food intake and occurs about every two hours. This juice is low in digestive power but important for the normal functioning of the bowels, while the second juice is stimulated by the presence of chyme in the duodenum and has mild digestive power. This second juice is mildly alkaline and contains peptidases, carbohydrases, phosphatases, nucleases, and enterokinases.

Enterokinase is the important kinase that activates the inactive chemical on the proper disaccharides as follows:

$$\text{sucrose + water} \xrightarrow{\text{sucrase}} \text{glucose + fructose at a 5.0-7.0 pH}$$

$$\text{maltose + water} \xrightarrow{\text{maltase}} \text{glucose + glucose at 7.0 pH}$$

$$\text{lactose + water} \xrightarrow{\text{lactase}} \text{glucose + galactose at 5.4-6 pH}$$

The concentration of lactase is greatest in the intestinal juices of young mammals.

The peptidases hydrolyze polypeptides to free amino acids; the peptidases are most active at 7-8 pH.

The nucleases and phosphatase are responsible for the hydrolysis of nucleoproteins into their respective units. Phosphatase is most active at a pH of 8.6 and can function properly in the duodenum.

The nutrients taken into the body are now ready to be absorbed and circulated to those parts of the body that need the materials.

(Use the chart Kinds of Digestive Enzymes in Figure 13-1 as a summary.)

KINDS OF DIGESTIVE ENZYMES

ENZYMES	SUBSTRATE	PRODUCT	COMMENT
CARBOHYDRASES			
Amylase	starch	maltose	Salivary amylase is frequently known by an older name, ptyalin.
Maltase	maltose	glucose	Wide occurrence
Cellulase	cellulose	cellose (a double sugar), also known as cellobiose	Of rare occurrence; in symbiotic protists, teredo, and silverfish
Hyaluronidase	hyaluronic acid, a carbohydrate, which "cements" cells of animal tissues together	shortened carbo-hydrate chains	Occurs in bacteria, snake venom, and testes of mammals. (Enzyme produced by testes enables sperm to penetrate cells surrounding egg.)
LIPASES			Of general occurrence in animals, plants, molds, bacteria
Gastric lipase	fats and related compounds	fatty acids and glycerol	Secreted by glandular tissue of stomach wall
Pancreatic lipase	fats and related compounds	fatty acids and glycerol	Produced in pancreas; enters small intestine through pancreatic duct
Lecithinase	lecithin	lysolecithin, a com-pound very destructive to red blood cells	Occurs in the venom of snakes, scorpions, wasps, and in many plants, notably certain fungi
PROTEASES AND PEPTIDASES			Most are highly specific, operating on peptide linkage adjoining specific amino acids.
Pepsin	peptide linkages	shortened peptide chains and amino acids	Secreted by glandular tissue of stomach wall
Trypsin	peptide linkages	shortened peptide chains and amino acids	Secreted by pancreas; enters small intestine through pancreatic duct
Papain	peptide linkages	shortened peptide chains and amino acids	One of a series of proteases extracted from plants, in this case the papaya, often used as meat tenderizers
NUCLEASES			
Ribonuclease	RNA	shortened chains and single nucleotides	General occurrence
Deoxyribo-nuclease	DNA	shortened chains and single nucleotides	General occurrence

Figure 13-1 Digestive enzymes

Demonstration: Water in Food

Place a sample of a food into a dry test tube. Heat gently over a Bunsen flame. You will see drops of moisture condense on the sides of the test tube.

Demonstration: Minerals in Food

Heat a sample of food, such as bread or milk, in a test tube or evaporating dish until it forms an ash or until the liquid has evaporated. A whitish ash indicates the presence of minerals.

Demonstration: Action of Pepsin

Make a solution of artificial gastric fluid by mixing 2 grams of commercial pepsin in 0.4% hydrochloric acid solution to make 100 ml of fluid. Filter to get a clear fluid. Test the resulting solution with litmus. Now number six test tubes from one to six and place in a rack. Into #1 put a few pieces of cracker; into #2 place a small piece of hard-boiled egg; into #3 place a small quantity of butter or margarine; into #4, #5, and #6, put a piece of the same food.

Into #1, #2 and #3, place about 1/3 of a test tube of the artificial gastric fluid, while in #4, #5, and #6, place about 1/3 of a test tube of plain tap water. Place all six test tubes in an incubator overnight at 37° C. The following day you will notice that the pepsin only acted on the protein egg white.

Demonstration: Digestion in the Mouth

(1) Place about one inch of starch suspension in each of two vials. Now take a microscope slide and place it on an overhead projector so that the entire class can see the demonstration. (2) Take a drop of starch suspension and place it on the left-hand corner of the slide. Add a drop of iodine to this drop and the

solution turns brownish. Add some saliva to each of the vials. Rotate the vials to mix the saliva with the starch suspension. (3) Select one vial and warm the starch suspension-saliva mixture by rotating between the hands. At the left end of your slide place a drop of the mixture, into which you place a drop of iodine. (4) Continue doing this at two-minute intervals until you have tested five drops, placing them from left to right on the slide. You will find that in time the addition of the iodine makes the color of the drop a deeper brown, until at the very end you have a deep brown color.

If you were to test the other vial at two minute intervals for sugar digestion by means of Benedict's solution, you would see that the Benedict's solution will turn green, but not red-orange, since there was not total digestion of starch in the mouth.

Demonstration: Action of Rennin

Rennin has a coagulating action. To show this, crush a Junket tablet and add it to a test tube of milk. Warm, but do not boil, the milk and you will notice how rapidly the milk solidifies in the test tube. However, if you were to boil the milk first, there would be no reaction with the powder, since heat destroys enzyme activity.

Demonstration: Role of Bile Salts

Add 5 ml of water to one test tube; to another tube, add 5 ml of bile suspension. Then sprinkle powdered sulfur on the surface of each test tube. The tube with the bile salts will allow the sulfur powder to fall into the test tube. To show what the value of reduced surface tension means on the emulsification of fats, take a test tube and add a 5% solution of bile salts, such as sodium taurocholate, and then add a few drops of olive oil. Shake well and allow to stand. There will be a permanent emulsion. There also will be very small drops of fat, which is a preliminary step in fat digestion by lipase in the small intestine.

Laboratory Exercise: Intracellular Digestion in Paramecium

Aim: To observe ingestion and digestion in paramecium.

Not being able to synthesize their organic compounds from simple inorganic materials, heterotrophs take in preformed organic compounds. These may be digested intracellularly, as in the paramecium. In this exercise you will observe how a paramecium ingests yeast cells and you will note changes inside the food vacuoles during the process of digestion.

Preparation for Laboratory: The yeast cells which will be fed to the paramecium have been prepared by boiling them in water to kill the cells so that their membranes will be permeable to the indicator, Congo red. Congo red is an indicator that is red at pH 5 and blue at pH 3.

Procedure: (1) Make a ring of methyl cellulose, about 3/4 inch in diameter, in the center of a microscope slide. Place a drop of paramecium culture inside the ring and add a cover slip. Focus with care in the usual manner with low and then high power on your microscope. The syrupy methyl cellulose will soon diffuse into the water and slow down the organisms. Observe their general structure and movement. (2) Now remove the cover slip and add a drop of suspension containing the dyed yeast cells. Cover and observe immediately. You should note the food vacuoles containing colored red yeast cells as the food is being swept into the body of the organism by its cilia. Follow the path of a food vacuole for approximately five minutes while looking for changes in the color of the yeast cells. As the food is being digested, the yeast cells will slowly change from red to blue, and you will be able to follow the movement of the yeast. You will note that the movement follows the regular movements of the protoplasm of the organism. Watch to see whether anything is being formed at the oral groove. If you watch long enough, you might notice material being egested from the paramecium.

Laboratory Exercise: Salivary Enzyme Activity and pH

Aim: To determine which pH is best for the salivary enzyme *amylase* to function at its maximum.

In this exercise you will determine what the pH will be for the maximum efficiency of amylase in saliva to start the breakdown of starch to maltose.

Materials Needed: Test tubes; test tube rack; 25 ml graduated cylinder; small beakers; spot plates; medicine dropper; flasks; starch solutions; iodine solution; buffer solution of pH 3, 5, 7, and 9; saliva (containing amylase); cheesecloth; marking pencil; sterile rubber bands or sterile paraffin; Benedict's solution; Bunsen burner; test tube holder.

Procedure: (1) Chew on rubber bands or a piece of paraffin until the saliva flows. Collect about 1/4 of a test tube full. Add an equal amount of water. (2) Shake the test tube thoroughly to mix and then filter through a double thickness of cheesecloth into a beaker. (3) Make up the pH solution of 3, 5, 7, and 9 by using the buffer tablets—one dissolved in 100 ml of distilled water. You can speed up the dissolving by crushing the tablet with a mortar and pestle. (4) When the buffer solutions are ready, take 5 ml of each and place each into a test tube. Label each test tube according to the pH of the solution. The buffer solution is a mixture of either a weak acid or base with one of its salts, which maintains the pH of the solution in which it has been placed. (5) Add 5 ml of starch solution to each of the test tubes and shake thoroughly to mix the contents. The starch is made by adding 5 gm of a soluble starch to 100 ml water. Boil and stir thoroughly to dissolve the starch. Allow to cool before using. (6) Add five drops of the salivary solution to each tube and shake thoroughly. Test each tube for the presence of starch by taking a few drops of the starch-saliva solution and testing with the iodine solution. (7) Continue to test the contents of each tube at one-minute intervals. As the starch changes to dextrin, the color becomes brown. When it remains colorless, it means that the substance now in the test tube is maltose. (8) To check for the presence of the maltose, use the Benedict's solution for a sugar test. A red-brown precipitate means sugar is present. (9) You should find that the starch is best changed into dextrins and then sugar at a pH of 7, with the next best being a pH of 5. The pH of 3 still shows starch present at the end of fifteen minutes, as does a pH of 9.

Laboratory Exercise: Gastric Digestion

Aim: To demonstrate the effect of pH upon the digestion of protein by pepsin.

Since it is necessary that there be a specific pH in the stomach for proper digestion, we shall attempt to find out what this pH should be. We are using egg white coagulum for the protein.

Materials Needed: (At the front desk) finger bowl containing raw egg white; capillary tubing, six pieces, each about eight inches long; Erlenmeyer flask or pot, eight inches deep; tripod; Bunsen burner; four triangular files and ampule knives. (For each group of students) test tube rack or glass tumbler for holding test tubes; four test tubes; 25 ml graduated cylinder; labels; bottles, each containing as follows: (1) HCl, 0.8% (2) pepsin, 2.0% (3) NaHCO3 0.8%, (4) water.

Procedure:

(A) Preparation of Solution:

One member of each group labels the four test tubes by recording the class, group and tube number, e.g., Biology 106, 3, IV.

A different student in each group prepares one of the four solutions. Wash the graduate each time it is used. *Solution I* will be 5 ml of HCl (0.8%) and 5 ml water. *Solution II* will be 5 ml of pepsin (2.0%) and 5 ml of water. *Solution III* will be 5 ml of pepsin (2.0%) and 5 ml of HCl (0.8%). *Solution IV* will be 5 ml of pepsin (2.0%) and 5 ml of NaHCO3 (0.8%).

(B) Preparation of Egg-White Tubes

The protein used is coagulated egg white, prepared from a fresh egg. One member of each group goes to the front demonstration table and draws egg white into a piece of glass tubing about eight inches long. The albumin is then coagulated by placing the tubes into water at 85° C for five minutes. The tubes are then cut into sections about one inch long. Each section is placed into each of the four test tubes, which meanwhile have been prepared by the other students in the group.

(C) Incubation:

Place all tubes into a glass tumbler which will fit into the incubator. Incubate for twenty-four hours at 37° C.

Observation: The tumblers containing the tubes should be examined the following day. Test tube #3 should show the most digestion.

Laboratory Exercise: Digestion in the Small Intestine

Aim: To learn what digestive enzymes are found in the small intestine.

The final steps in digestion are carried out in the small intestine. Fats are changed into fatty acids and proteins into amino acids. Pancreatin contains the digestive enzymes used in these changes.

Materials Needed: Pancreatin powder; chopped hard-boiled egg; eleven test tubes; phenolphthalein solution; 0.5% sodium carbonate solution; 2% boric acid solution; water; olive oil; soap solution or 5% bile salt solution (sodium taurocholate); dried milk powder; litmus powder.

Procedure:

(A) Digestion by Trypsin:

(1) Put some chopped hard-boiled egg white into six test tubes; add a pinch of pancreatin along with 10 ml of water. (2) Set two tubes aside. (3) To two others, add two drops of phenolphthalein solution. Add 0.5% solution of sodium carbonate, drop by drop, until the first pink color appears. The pH of these tubes will be approximately 8, or on the alkaline side. (4) To the last two tubes add 5 ml of a 2% boric acid solution so that they will contain a pH of close to 5. Place all the test tubes into a water bath at 40° C or an incubator for twenty-four hours. The most digestion will be found in those tubes that were slightly alkaline along with the pancreatin.

(B) Formation of Acid:

Pancreatin contains digestive enzymes which split fats into fatty acids and proteins into amino acids. In order to show their reaction, litmus is used as an indicator to show that acids were formed after digestion. (1) Prepare litmus milk powder by adding one part of litmus powder to forty parts of dried milk powder. Then dissolve one part of the prepared litmus powder in nine parts of water. (2) In each of two test tubes place 10 ml of the solution. Then place a pinch of pancreatin in one and to the other add only water. Place the tubes in a water bath. (3) There will shortly be a change of the litmus solution to colorless. This indicates that the milk has broken down to an acid.

(C) Breakdown of Fats:

(1) Prepare three test tubes containing 10 ml of water and a few drops of olive oil. (2) Add 2 ml of a soap solution to two of them. (3) Add a pinch of pancreatin powder to 10 ml of water and place 5 ml of this solution in one of the test tubes with the soap solution. You now have three test tubes: (a) a tube containing water, olive oil, soap solution, and pancreatin solution; (b) a control containing water, olive oil, and soap solution; (c) another control containing water, olive oil, and pancreatin water. (4) Use litmus paper to show that the solutions are neutral to start with. Place the test tubes into a water bath for thirty minutes. (5) Tube (a) will have acid in it when tested with litmus after thirty minutes.

AUDIO-VISUAL MATERIALS

FILMS

Chemical Digestion (B)
The Digestive System (K)
Food and Nutrition (B)
Nutrition and Metabolism (K)

FILMSTRIP

Digestion (The Living Cell Set) (A)

TRANSPARENCY

Digestion and Mechanism of Enzyme Action (KK)

KIT

Food Analysis Kit (F), (O), (CC), (EE), (FF)

Vitamins, Hormones and Blood

UNIT 14

These areas of biology are fairly small; therefore they will be placed together into one unit for simplicity.

VITAMINS

The first of these three areas in the field of biology is vitamins. We shall not go into the general background and history of vitamins, as you have gone into that before.

Very early in the history of life it became apparent that all life forms depend for their existence upon one another: oxygen, necessary for the life of many animals, was supplied by many plants in return for carbon dioxide, given off by many animals; plants themselves became food for some animals that, in turn, became food for other animals, etc. So, too, did "growth factors" fit into this scheme of the interdependency of life: certain plants produced growth factors that were necessary for the life of other plants and animals—and could be obtained only through food by eating the life that produced the necessary growth factor.

In 1912, Funk, after associating several diseases with a deficiency in the diet of certain of these growth factors, called these factors "vitamines" from the Latin word *vita* meaning life and *amine* an organic chemical designated as the organic base.

Today, many of these "vitamines," now shortened to "vitamins," have been isolated, some of their functions have been described, and a number of them have been made synthetically.

Biochemists continued to work on the chemical structure of vitamins and now know quite accurately the chemical structure of about fourteen vitamins. Many have turned out to be groups of related vitamins rather than single substances. Some are referred to by chemical names, such as thiamine and niacine, while others are still referred to by letter, such as D and K. But we still classify vitamins as fat-soluble or water-soluble.

Vitamins are organic compounds and do not resemble each other except that vitamins are organic compounds, required in small amounts, essential in metabolism and therefore necessary.

Vitamins play an essential role in the production of many chemicals that are needed by the body to carry out its various functions of growth, reproduction, respiration and movements.

Vitamin A

For more than 4,000 years man has known that certain foods, such as raw liver and fish liver oil, were effective in the correction of night blindness and the dry, lusterless condition of the eyeball now known as xerophthalmia. However, it was not until 1913 that scientists realized for the first time that a hitherto unsuspected factor present in foods was actually responsible for the correction of the conditions. This factor, a fat-soluble chemical, is known as Vitamin A.

Vitamin A is classified as an alcohol of high molecular weight (286.44), having an empirical formula of $C_{20}H_{30}O$ which is described as 3, 7-dimethyl-9-(2, 6, 6-trimethyl-l-cyclohex en-1-yl)-2, 4, 6, 8, -nontetraen-1-ol. Its structural formula is:

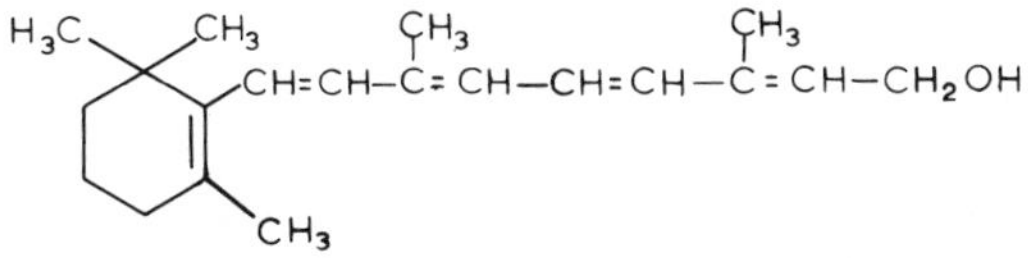

Vitamin A is found naturally in animal products only. The high vitamin A activity of green leaves, green and yellow seeds, and various fruits and roots having yellow carotenoid pigment is due to the chemical precursors to vitamin A, called carotenoids, that they contain. These carotenoid pigments are converted into vitamin A by an enzyme called carotenase, found in the liver of many animals.

Vitamin A is essential to maintaining the integrity of epithelial cells and stimulating new growth. It plays an important part in maintaining resistance to infections, and increases longevity. As an essential for the regeneration of "visual purple" it plays an important part in the sense of sight.

Excessive intake of vitamin A or its precursors may produce yellow discolorations of the skin, enlargement of the liver and spleen, drying or peeling of skin, loss of hair, nausea, headache, and blood dyscrasias.

Thiamine—Vitamin B-1

The symptoms of beriberi, which include weakness, pain and stiffness of the muscles, insomnia, paresis, and cardiac symptoms, have been known to man almost since the beginnings of civilization. In the East, particularly, it was prevalent, and it has been noted especially in armies, prisons, ships, and wherever large numbers of men are kept together.

By 1937 Williams had accomplished the synthesis of the vitamin, thus bringing to a successful conclusion the studies on the chemical nature of the first vitamin to be definitely postulated.

Thiamine is usually sold as the hydrochloride or the mononitrate. As the hydrochloride it has a molecular weight of 337.28 and an empirical formula of $C_{12}H_{17}ClN_4OS \cdot HCl$. It is described as 3-(4-amino-2-methylpurimidyl-5-methyl)-4-methyl-5-hydroxy-ethyl-thiazolium chloride hydrochloride. Its structural formula is shown as:

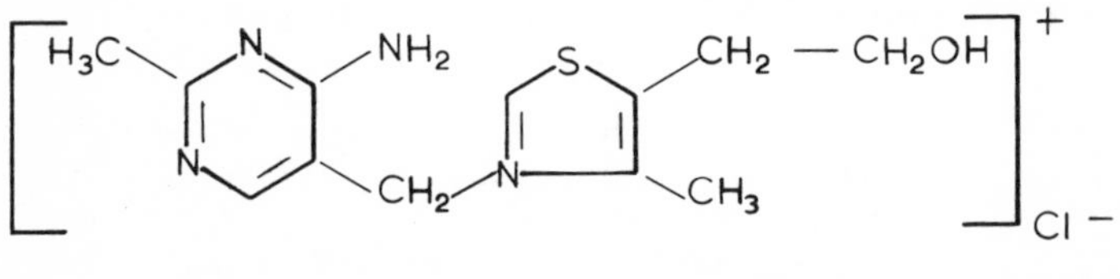

Vitamin B-1 is essential for maintenance of good appetite, normal digestion, and gastrointestinal tonus. It is needed for normal functioning of the nervous tissue (synthesis of acetylcholine) and for growth, fertility, and lactation.

Biochemically, its pyrophosphate ester, called cocarboxylations, leads to the formation of CO_2.

Riboflavin—Vitamin B-2

Riboflavin is a water-soluble, yellow pigment that has a molecular weight of 376.36 and an empirical formula of $C_{17}H_{20}N_4O_6$. It is described as 6, 7-dimethyl-9-(D-l'ribityl) isoalloxazine. Riboflavin has the structural formula:

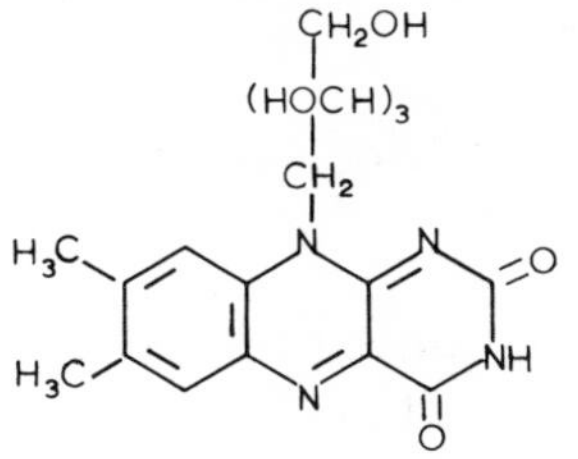

Riboflavin, as riboflavin-5-phosphate, is a component of a number of essential enzymes concerned with the oxidation and reduction of body chemicals.

Nicotinic Acid (Niacin, Nicotinamide, Niacinamide)—P.O. Factor

Pellagra has been known for centuries, but it wasn't until 1914 that Voegtlin proved, without question, that the disease was caused by a dietary deficiency. In 1935, sixty-eight years after Huber described crystalline nicotinic acid, this chemical and its amide were conclusively identified as a component of an enzyme system indispensable for health.

Niacin, a water-soluble vitamin, has a molecular weight of 123.11, an empirical formula of $C_6H_5NO_2$, and is chemically known as pyridine-3-carboxylic acid. Its structural formula is:

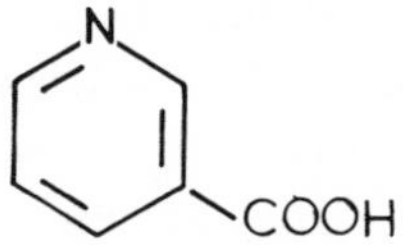

Niacinamide, also water-soluble, has a molecular weight of 122.12 and an empirical formula of $C_6H_6N_2O$. Its structural formula is:

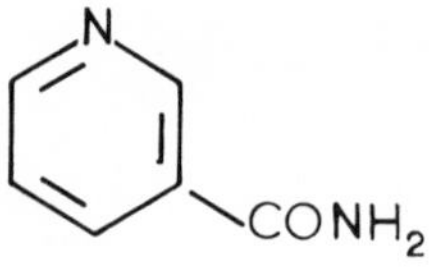

Niacinamide is a functional group of coenzymes that is concerned with dehydrogenation, especially in carbohydrate metabolism.

Pyridoxine, Pyridoxal, Pyridoxamine —Vitamin B-6

In 1934, Gyorgyi proved the existence of a vitamin essential for rats. Pyridoxine hydrochloride has a molecular weight of 305.64, an empirical formula of $C_8H_{11}NO_3 \cdot HCl$, and is also known as 5-hydroxy-60methyl-3,4-pyridine-dimethanol hydrochloride. Its structural formula is:

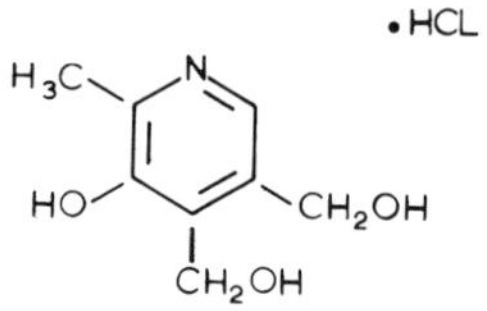

It is a water-soluble vitamin that is essential for the complete metabolism of tryptophane and is contained in the coenzyme for amino acid decarboxylase and transaminase.

Pantothenic Acid

In 1939, Williams reported an acidic substance which was required as a growth factor for yeast. Because this substance has such a universal distribution in biological material it was named "pantothenic acid" which means, literally, "from everywhere."

Pantothenic acid, a water-soluble vitamin, has a molecular weight of 219.23, an empirical formula of $C_9H_{17}NO_5$, and is described as D(+)-N-(a,-Dihydroxy-b-b-dimethylbutryl)-b-alanine. Its structural formula is:

$$HOCH_2-\underset{\displaystyle CH_3}{\overset{\displaystyle CH_3}{\underset{|}{\overset{|}{C}}}}-CHOH-CONH-CH_2-CH_2-COOH$$

Pantothenic acid is essential to life. It is a component of coenzymes and is concerned in some way with fat and carbohydrate metabolism. It is related to the utilization of other vitamins, especially riboflavin, and is believed to be a growth factor necessary to promote vitamin synthesis by the intestinal flora.

Folic Acid—Vitamin B

In the elementary studies of nutritional requirements of microorganisms, as early as 1937 several scientists realized that *Lactobacillus casei* required a substance for its growth that was present in green leaves, grasses, etc. It later proved useful in curing and preventing certain laboratory-induced anemias. It was later synthesized and identified as pteroylglutamic acid.

Folic acid is a water-soluble nutritional factor that has a molecular weight of 441.40, an empirical formula of $C_{19}H_{19}N_7O_6$, and is known as pteroylglutamic acid and N-[4-[(2-amino-4-hydroxy-6-pteridyl-methyl]-amino]-benzoyl]-glutamic acid. Its structural formula is:

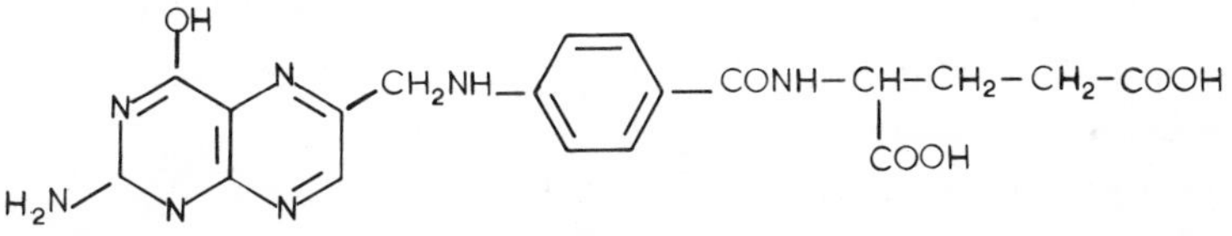

Folic acid is one of the more fundamental molecules essential to the normal metabolism of all types of cells in the marrow, as well as many other growing cells and tissues.

Cyanocobalamin—Vitamin B-12

Cyanocobalamin is a water-soluble vitamin having a molecular weight of 1357 and an empirical formula of $C_{62}H_{90}O_{14}N_{14}PCo$, being the only vitamin containing a mineral atom. Its structural formula follows.

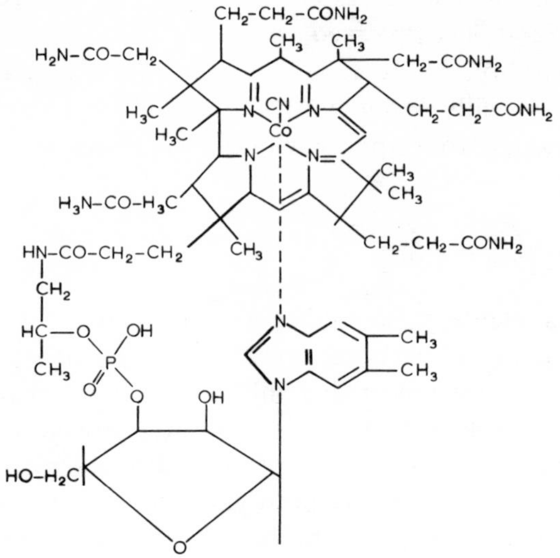

Ascorbic Acid—Vitamin C

Scurvy has been known to seafaring men almost since the beginning of time. Drinking lime juice was part of a dietary reform for scurvy.

In 1928, Gyorgyi isolated a substance from adrenal glands, cabbage and oranges that he called "hexuronic acid," which was later proved to be identical to Vitamin C, or as it was renamed, in 1933, ascorbic acid. Without knowing the molecular structure of this vitamin, Reichstein, Grussner, and Oppenheimer synthesized it in 1933.

Ascorbic acid is a water-soluble vitamin that has a molecular weight of 176.12, an empirical formula of $C_6H_8O_6$, and a structural formula of:

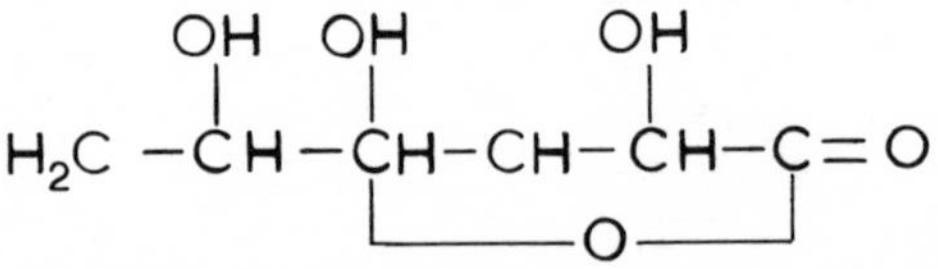

Ascorbic acid occurs in a wide range of animals and plants, in fact, all plants and most animals can synthesize their own Vitamin C.

Vitamin C's primary function seems to be in enhancing the production of intracellular materials of connecting and supporting tissue. In this way it aids in the formation of bones and repair of fractures, the healing of wounds and the prevention of capillary hemorrhage.

Calciferol—Vitamin D

Calciferol, a fat-soluble vitamin prepared by irradiation of ergosterol, has a molecular weight of 296.63 and an empirical formula of $C_{28}H_{44}O$. Its structural formula is:

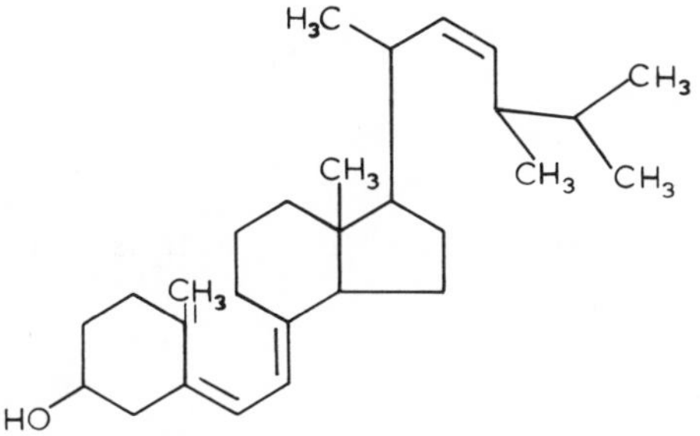

Minor differences in structure result in D-2, D-3 or D-4.

Vitamin D enhances the absorption of calcium from the intestinal tract, promotes the conversion of inorganic phosphorus to organic forms in the bone, and, in general, concerns itself with the formation of bones and teeth.

Alpha-Tocopherol—Vitamin E

Alpha-tocopherol is an oil- or fat-soluble vitamin having a molecular weight of 430.69, an empirical formula of $C_{29}H_{50}O_2$, and is known as 5, 7, 8-trimethyltocol. Its structural formula is:

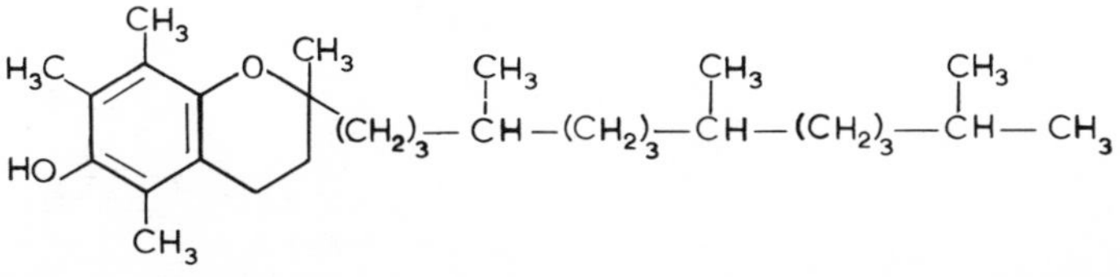

Alpha-tocopherol is necessary for normal reproduction in many animal species, but its significance in human nutrition has not been established. It may, however, act as a regulator of the metabolism of the cell nucleus, especially concerned with cell maturation and differentiation.

Menadione—Vitamin K

In nature there are several vitamin Ks, but menadione, one of the more popular synthetic vitamin Ks, has a molecular weight of 172.17 and an empirical formula of $C_{11}H_8O_2$. It is also known as 2-methyl-1-4-naphthoquinone. Its structural formula is:

It is known as a fat-soluble vitamin.

Vitamin K is essential for the synthesis of prothrombin and normal blood clotting.

Biotin

Biotin has an unusual structure and is known as 2'-keto-3-,4-imidazolido-2-tetrahydrothiophene-n-valeric acid. Its structural formula is shown.

Choline

Choline is a derivative of ammonium hydroxide. It is the nitrogenous base in lecithins and spingomyelins, and a lipotropic factor in the metabolism of lipids in the liver. A deficiency of choline in the diet reduces the phospholipid content of the liver, causing the piling up of fat in the organ and impairing its function in fat transport. It is a supplier of methyl groups for transmethylation reactions. Its methyl groups are available for the body to use in the synthesis of creatine, creatinine, adrenaline, and several other compounds. It is also important in the transmission of nerve impulses in the form of its ester, acetycholine.

HORMONES

Hormones are chemical messengers which help in controlling and coordinating the activities of the body. They are produced in the endocrine or ductless glands, which the students have heard of. The ductless glands pour their secretions directly into the bloodstream. The hormones are produced in one area and are transported to another, where they exert their influence. They are effective at extremely low concentrations and some act more effectively if present in combination with another. Others are produced more or less continuously, while many are produced in response to some environmental condition. Their concentration is controlled in part by the liver, which provides for their destruction. A normal balance of their metabolic activity is maintained at all times in a healthy organism.

Some hormones are proteins, others are polypeptides or steroids. Various hormones, in their function of coordinating the responses of the body, interact in rather complex ways. They interact with other chemical agents, called tissue hormones or parahormones. For example, carbon dioxide, made by all cells, stimulates the respirator and other brain centers, while other tissue hormones are vitamin F, serotonin, histamine, and acetylcholine.

(There are no demonstrations or laboratory exercises for this part of the unit since there is little that can be shown that would have a bearing on the biochemical angle of hormones.)

In research with arthropods, it has been found that endocrines are important in the metamorphic changes that the insect goes through. It has also been shown that endocrines function in the reproductive activities and color changes which some species of insects undergo. Hormones also regulate molting and color change in crustaceans. Among the vertebrates, research has shown fishes, amphibians, reptiles, and birds have endocrine glands similar to those of mammals, even in respect to location and hormone secretion. The action of specific hormones in some members of these classes may be more prominent in one species than in another.

Integration of the entire endocrine system is centered in the pituitary gland or "master gland." The pituitary consists of three distinct regions, each of which secretes distinct hormones. The pituitary serves as a central control panel for the entire endocrine system.

The thyroid gland has no activity under normal conditions, but goes into action when stimulated by the thyrotopic hormone of the pituitary. When active, the cells of the thyroid produce the amino acid called thyroxin, which includes four atoms of iodine. After its secretion into the bloodstream, thyroxin is bound to a large protein, a globulin. Thyroxin accelerates the utilization of all energy foods. It also has other effects.

Imbedded in the thyroid are four parathyroid glands. The bioregulator is parathormone, which regulates both calcium and phosphorus metabolism.

The pancreas has an endocrine function in addition to its digestive activity. It produces insulin, which regulates the amount of sugar in the blood.

The sex hormones are responsible for the normal growth and development of the secondary sex characteristics of most vertebrates. The gonads, which produce the sex hormones, are also responsible for the production and development of the sex cells, or gametes.

The adrenal glands are two small structures situated bilaterally just above the kidneys and produce a number of important hormones, such as adrenaline which has a role in the operation of the nervous system. Other adrenal hormones regulate the metabolism of proteins, fats, and carbohydrates, while some regulate the movement of sodium and potassium ions and of water.

Secretin is liberated into the bloodstream from the lining cells of the duodenum when the cells are stimulated by a decrease in pH. The digestive enzymes of the pancreas empty into the small intestine only when food is present there. Secretin also stimulates the production of the bile in the liver. Another gastrointestinal hormone controls the emptying of the gall bladder.

(The chart *Hormones and Their Effects* (Figure 14-1) summarizes the information needed on hormones.)

GLAND	HORMONE	SITE OF ACTION	EFFECT
Pituitary	Growth hormone	General	Increases growth
	Thyroid-stimulating hormone (TSH)	Thyroid	Increases thyroid hormone production
	Adrenocorticotropin (ACTH)	Adrenal cortex	Increases cortical hormone production
	Luteinizing hormone (LH)	Ovary	Increases progesterone
		Testis	Increases testosterone
	Follicle-stimulating hormone (FSH)	Ovary	Stimulates ovulation; increases estrogen
		Testis	Increases sperm production
Thyroid	Thyroxin	General	Increases metabolic rate
Parathyroid	Parathormone	Bone and kidney	Regulates calcium metabolism
Testis	Testosterone	Accessory sex organs	Stimulates function
		General	Stimulates development of male characteristics
Ovary	Estrogens	Accessory sex organs	Stimulates normal functioning
		Uterus	Prepares wall for pregnancy
		General	Stimulates development of female characteristics
	Progesterone	Uterus	Maintains wall
Adrenal medulla	Adrenaline	Heart and smooth muscle	Increases pulse and blood pressure
		Liver and muscle	Controls breakdown of glycogen
Adrenal cortex	Corticosterone	General	Increases metabolic rate; affects sensitivity
Pancreas	Insulin	General	Reduces blood sugar level
Lining cells of duodenum	Secretin	Pancreas and liver	Releases pancreatic juice and bile
Nerve cells	Acetylcholine	Synapses and nerve-muscle junction	Reduces electrical resistance

Figure 14-1 Hormones and their effects

BLOOD

We now will cover the third topic of this unit, which is familiar enough to you and the students, but contains a number of facts that might be rather vague biochemically.

The blood animals have is called the circulating or "floating" tissue of the body because of the great number of cells and substances found in its solution. It has several important functions: (1) carries oxygen from the lungs or breathing apparatus and nutrients from the intestinal tract to the body cells; (2) carries the waste products of metabolism from the body cells to the organs of excretion; (3) maintains the pH of the body by means of its buffer compounds; (4) governs water balance and fluid distribution to a large extent by the inorganic salts and proteins present in the blood; (5) carries antibodies and white blood cells, which help to protect the body against invasion by bacteria and viruses; (6) carries detoxified products and various end products of metabolism, such as CO_2 and urea, to the proper organs and eliminates them; (7) carries hormones from ductless glands to sensitive tissues; (8) serves as a heat exchange, preventing extremes in body temperature in warm-blooded animals.

The blood of higher animals is made up of red blood cells, white blood cells, and platelets suspended in a straw-colored fluid called plasma. Whole blood contains about twenty percent total solids, about half of which is the chromoprotein *hemoglobin,* and most of the remainder is plasma protein, which is mainly albumin and globulin. The formed elements are in the ratio of about 1:100:1000 of white cells; platelets; red cells. Adult human females have a slightly lower red cell count than that of the male. Red cells and white cells may exist in several different shapes. A normal platelet level is essential for proper clotting of the blood.

Blood plasma contains more than fifty different components, most of which are in trace amounts. The most abundant components are albumins (about five percent) and the globulins (about two percent). Albumins serve as an important function as colloids for the maintenance of the proper exchange of fluid between blood and tissues. Alpha and beta globulins have the same functions, and gamma globulins consist mainly of antibodies for certain bacteria and viruses. The trace proteins, fibrinogen, prothrombin, and some activating globulins, are important factors in blood coagulation. Glucose, amino acids, lipids, mineral ions, vitamins, and oxygen, held in loose chemical combination by hemoglobin and carbon dioxide, also are to be found in chemical combination in blood plasma.

For defense, the body depends on both plasma proteins and cells. The plasma contains a special group of proteins, called antibodies, which combine with and hence inactivate foreign proteins, viruses, or polysaccharides, and also cause invading bacteria to clump together. Each antibody is specific for the substance or type of cell with which it reacts. Somehow our defense machinery knows the shapes of our own proteins and leaves them alone. When foreign proteins or polysaccharides called antigens are introduced in the circulation, antibodies against them are quickly synthesized. A specialized group of white blood cells, the plasm cells (plasmocytes), produce antibodies.

Blood also provides a constant internal environment for the cells and tissues of the body. In a mammal the pH, temperature and sugar concentration of the blood are held within very narrow limits. This relative stability of the internal environment makes it possible for a mammal to experience enormous changes in the external environment without damage.

The many vital functions performed by blood and its components make it necessary that the blood be maintained at a constant composition and volume in the closed vessel system. The coagulation of blood is an example of delicately balanced protein interactions. The coagulation of blood is divided into two major steps, the conversion of plasma prothrombin to thrombin and the conversion of plasma fibrinogen to fibrin. Prothrombin is a proenzyme containing within its structure two enzymes, thrombin and autoprothrombin C. Both of these enzymes are capable of catalyzing the conversion of prothrombin to thrombin. Thrombin and autoprothrombin are glycoproteins and thromboplastin is a complex that contains protein, lipid, nucleic acid, and carbohydrate.

Lymph circulates slowly in the lympathic vessels and is returned to the blood via special lympathic ducts. The composition of lymph is essentially the same as plasma, except that the level of proteins is slightly lower and the levels of nitrogenous waste products are slightly higher. Lymph is relatively free of formed elements, and its coagulation time is therefore much slower than that of whole blood.

(There are several demonstrations you can do that deal with coagulation and precipitation in blood.)

Immunity

Immunity to infection is dependent on the action of the white blood cell (leucocyte) and the action of the plasma antibodies. The efficiency with which the phagocytic white blood cells engulf and digest bacteria depends to a great extent on the content of specific substances (probably proteins) of the plasma, called opsonins.

It is possible that any protein which is foreign to the animal or human body can act as an antigen, and when injected will call forth an antibody response. Antibodies are: (1) agglutinins, that can cause clumping together of cells carrying the antigen; (2) antitoxins, which neutralize the toxin released by a specific pathogenic organism; (3) hemolysin and other cytolysins, which destroy cell membranes and cause liquefaction of the cell; (4) precipitins, which form an insoluble compound specifically with the antigenic protein. For the antibody to react with its antigen, an additional factor, called a "complement," is needed. This complement is a complex globulin type of protein.

When one is sensitive to certain proteins he is considered to have an allergy toward that particular substance. It is a form of anaphylaxis or hypersensitivity to traces of foreign protein. Desensitization to allergens is accomplished by the repeated injection of the allergen in small doses, so that the body can build protective antibodies instead of sensitizing antibodies.

Hemoglobin

Hemoglobin is a crystalline protein with a total molecular weight of about 65,000. It is composed of four heme groups to each globulin part. Hemoglobin readily combines with oxygen to form oxyhemoglobin, bright red arterial blood. Hemoglobin is purplish red, venous blood. The iron in the heme part is in the ferrous state.

The body is constantly creating new red cells and destroying old red cells. The heme part (minus the iron) is called the protoporphyrin part; it can be oxidized and reduced, and is excreted as bile pigments in the feces and urine.

Since hemoglobin contains heme and proteins, various species have characteristic proteins and the hemoglobin for each species is different.

Blood Types

The red blood cells contain two substances capable of being precipitated or clumped, and they are referred to as agglutinogens A and B. The agglutinogens are protein in nature and may be regarded as antigens. You may have either, both, or neither of these substances. If you do not have agglutinogens in the red blood cells, your blood type is O; you have type A if the agglutinogen in your red blood cells is A; you have type B if the agglutinogen in your red blood cells is B; and your blood type is AB if you have A and B agglutinogens in your red blood cells. When blood from individuals of different blood types is mixed, agglutination of the red blood cells of one (or both) of the bloods takes place. It is necessary, therefore, to use the patient's type of blood when he receives a blood transfusion. A mixture of two different blood types in a person will cause the antigens to clump together or agglutinate and may kill the person. It is the antigen more than the antibody of the donor which causes severe damage when injected into an incompatible person, since the bulk of the blood cells agglutinated are those of the donor. (Do the blood typing as a laboratory exercise.)

Rh Factor

Recently a number of blood factors in addition to the blood types have been discovered, and the most important is the Rh factor. About 85 percent of the white race and 99 percent of certain other groups have an antigen called the Rh factor in their red blood cells. Blood containing the Rh factor is known as Rh positive; that lacking it, Rh negative. If blood from an Rh positive person is transferred into an Rh negative recipient, antibodies (anti-Rh factor) are produced. This, in itself, is not harmful, but if a second transfusion is given, the anti-Rh antibody which has accumulated in the recipient's blood reacts with the Rh antigen introduced with the new red blood cells, and the results are often fatal.

The anti-Rh antibody may also be produced in an Rh negative woman who, having an Rh positive husband, bears Rh positive children. During pregnancy, some fetal red blood cells containing the Rh antigen may leak into the mother's circulation and cause the formation of the anti-Rh antibody. This has no ill consequences for the mother unless she later receives a transfusion of the Rh positive blood. Usually the first child is not seriously harmed, since the mother's antibody titer is still low. During later pregnancies, however, the mother's anti-Rh antibody may enter the fetal circulation and destroy the fetal blood cells. It can cause the death of the child, unless its blood can be replaced in a massive transfusion by Rh positive blood free from the antibody.

pH

The ph of the blood must be maintained within narrow limits or serious effects are observed. The pH range of the blood is normally 7.35-7.45. Symptoms of acidosis are observed if the pH falls below 7.35, with the death of the animal at a pH of about 7; alkalosis is observed if the pH rises to about 7.45, and the animal dies at a pH of about 7.8.

Demonstration or Laboratory Exercise:
Test for Ascorbic Acid

Aim: Ascorbic acid acts as a reducing agent and has the ability to reduce certain chemicals. In this exercise you will learn how a starch-iodine solution can be used as a test for ascorbic acid.

Materials Needed: Beaker, glass rod, cornstarch, tripod, iodine solution, Bunsen burner, seven test tubes, test tube rack, 10% ascorbic acid solution, 1% ascorbic acid solution, grapefruit juice.

Procedure:

(A) Preparation of Starch-Iodine Solution

(1) Measure 100 ml of water into a beaker and heat to boiling point. Remove from heat. (2) Measure one gram of cornstarch; stir into warm water until dissolved. Cool the solution. (3) Measure 10 ml of the starch suspension into each of seven test tubes. Number the test tubes and put a drop of iodine into each tube. Shake well; the resulting solution should be blue in color.

(B) Ascorbic Acid and Starch-Iodine Solution

(1) Add 10% ascorbic acid a drop at a time to tube #1. Shake well between each drop. Count the number of drops required to remove the blue color. It should be about three. (2) To test tube #2, add the 1% ascorbic acid drop by drop and shake between each drop. Record number of drops required to remove the blue color. It should be about six, showing that a weaker ascorbic acid has a lower bleaching ability.

(C) Starch-Iodine Solution as a Test for Vitamin C

(1) Add the grapefruit juice a drop at a time to one tube of starch-iodine suspension and shake between each drop until the blue color is bleached out. The number of drops will depend upon the juice used. (2) Now boil a sample of the juice and allow it to cool. Test this juice for vitamin C by adding it a drop at a time to another test tube with the starch-iodine suspension. There will be more drops needed. (3) Now take a sample of the juice and add two pinches of sodium bicarbonate. Shake well and test this juice by adding it, a drop at a time, to the last test tube of starch-iodine suspension. There will hardly be any bleaching because the acid has been neutralized by the sodium bicarbonate.

Demonstration or Laboratory Exercise:
Use of Indophenol as Indicator for Vitamin C

Another solution can be used to show the bleaching power of ascorbic acid. This solution is 2,6 dichlorophenolindophenol, commonly known as "indophenol." You can do parts (B) and (C) of the preceding exercise for this exercise, using indophenol instead of the starch-iodine suspension. The results will be similar.

Laboratory Exercise: Homeostatic Condition of the Blood

Aim: To learn how the kidney maintains the homeostatic condition of the blood and to discover the role of the kidney in excretion.

Excretory organs are found in those animals that consume food containing large volumes of nitrogenous compounds, or have skin or scales or an exoskeleton that covers the living cells of the body surface so that the surface cannot act as an excretory surface. In the human, the kidney serves another purpose besides excretion. It is also used to maintain a constant level of chemical substances in the blood.

In order to show the effect of food intake on the composition of the urine, it is necessary that different students have different diets for one day prior to the laboratory test. There are five different diets: (1) Diet 1 is high protein: meat and other high-protein foods are eaten while starches, fats, and sugars are not. (2) Diet 2 is low-salt: drink plenty of water and make certain that the diet contains no salt of any kind. (3) Diet 3 is high glucose: eat more candy bars and less regular food; chocolate cakes and cookies are on the diet. (4) Diet 4 is high pentose: eat fruits, such as plums, pears, and bananas, instead of most of your regular diet. (5) Diet 5 is your regular diet: eat what you usually eat, which usually contains a little of everything.

Materials Needed: A stoppered bottle for urine sample, 50 test tubes and corks, glacial acetic acid, a methyl alcohol solution of xanthydrol, pipette, 10% silver nitrate solution, Clinitest paper, Bunsen burner, large beaker, graduated cylinder, Bial's reagent, 1.0% ferric chloride solution, 0.1M urea solution, 0.1% glucose solution, 0.1% sodium chloride solution, 0.25% pentose sugar solution.

Procedure: The class should work in groups of five, each member of the group choosing a different diet so that the entire set of tests can be done by each group and the information compared and exchanged.

(1) Each student should adhere to his particular diet for the day preceding the laboratory. On the morning of the laboratory, the student should collect his urine sample immediately upon awakening and label the bottle with the proper diet number.

(2) Each student will test his own urine with all four tests, as follows: (a) *Test 1. Urea:* Dilute 1 ml of urine with water to make 500 ml of solution. Transfer 5 ml of this diluted urine to a test tube and add 5 ml of glacial acetic acid and 0.5 ml of a methyl alcohol solution of xanthydrol. Cork the tube and shake it

vigorously. Into a second tube, place 5 ml of the urea solution found on your desk and add 5 ml of glacial acetic acid and 0.5 ml of the methyl alcohol solution of xanthydrol. Cork the bottle and shake it vigorously. Mark the tubes 1-a for the urine and 1-b for the urea tube, which is a control. Allow them to stand for at least one hour, overnight if necessary. If there are large, loose clumps in the tubes, urea was present.

(b) *Test 2. Chloride:* Dilute 1 ml of urine with water to make 500 ml. Transfer 5 ml of diluted urine to a test tube and add a few drops of 10% silver nitrate solution. To another tube add 0.1% sodium chloride solution plus a few drops of 10% silver nitrate solution. The second tube is the control and will show a cloudy white precipitate, indicating that a chloride is present. If there is a precipitate in the urine sample, there is chloride present in the sample.

(c) *Test 3. Glucose:* Dip a piece of Clinitest paper in the urine sample. Then compare the color of the paper with the chart provided with the paper to obtain the approximate percentage of glucose in the urine. Dip the paper into the glucose solution to compare the results of both.

(d) *Test 4. Pentose Sugar:* To 2 ml of urine in a test tube and to 2 ml of 0.25% pentose sugar solution in another test tube add 2 ml of Bial's reagent. *(Note: Corrosive. Be careful not to get any on your skin. If you do, wash under heavily running water.)* Gently warm the tubes over the Bunsen burner and then add a few drops of the 1.0% ferric chloride *(also corrosive).*

(3) The results should be tabulated on a separate sheet of paper and should generally show the following: (a) A high-protein diet will increase the amount of urea in the urine. (b) A salt-free diet should lower the amount of chloride in the urine. (c) A high-glucose diet may increase the amount of glucose in the urine. (d) A high-pentose diet will increase the amount of pentose in the urine. (e) A regular diet should not show any unusual amounts of any one substance in the urine.

Conclusion: We thus see that the amount of the various substances found in the urine depends upon the substances in the foods eaten. The kidney thus serves to maintain the homeostatic condition of the blood, selectively removing the excess substances from the blood and excreting them. Without the kidney functioning as it does, there would be a widely varied amount of dissolved substances in the blood, which would result in all sorts of physical malfunctions and illnesses.

Laboratory Exercise: Blood Typing

Aim: To determine your blood type and to see what happens when blood of different types is mixed.

Individuals may or may not have one or both of the known clumping factors in their red blood cells. The blood type of an individual will be determined by these clumping factors. There are two types of antibodies found in the blood serum which react against the clumping factors, but in any person, the antibodies of that person will never be antagonistic to his own clumping factors.

Blood Type	Clumping Factor in Red Blood Cells	Antibodies in Blood Serum
A	A	B
B	B	A
AB	AB	NONE
O	NONE	A and B

In this exercise you will be introduced briefly to the topic of blood typing by adding Anti-A serum and Anti-B serum to drops of blood and thus be able to determine the blood type.

Materials Needed: Anti-A serum (colored blue); Anti-B serum (colored yellow); toothpicks; slides; disposable sterile lancets; alcohol; absorbent cotton; glass-marking pencils; 0.85% salt solution and the seradiluted 1/4, 1/16, or oxalated blood.

Procedure:

(A) Typing Your Blood

(1) Prepare a clean slide by drawing a line down the center, marking the left side of the slide "anti-A" and the right side "anti-B." (2) To draw the blood, wipe the tip of your finger thoroughly with a piece of cotton soaked in alcohol. Jab the finger tip with the sterile lancet and then place the drop of blood on each side of the slide. Wipe your finger thoroughly with the

alcohol again. (3) Add one drop of the Anti-A serum (blue) to the A spot and one drop of the Anti-B serum (yellow) to the B spot. (4) Use a clean toothpick to mix the serum and blood in each circle. (5) Gently tilt the slide occasionally as you watch for three to five minutes. Look at both circles under the microscope. (6) Watch for clumping. Clumping in the A circle indicates the presence of A agglutinogens, and clumping in the B circle indicates the presence of B agglutinogens. Type A blood has A agglutinogens, Type B has B agglutinogens. Type AB has both, and Type O has neither. (7) What is your blood type?

(B) Effect of Mixing Various Bloods

(1) Work with one of your classmates on this part. On a clean slide place a drop of the salt solution. On one side of the salt solution you place a drop of your blood, while your classmate places a drop of his on the other side. Use a clean toothpick to mix the three drops. Slowly tilt the slide as before and wait for results. (2) If there is no clumping, it indicates that the two bloods are compatible. If there is clumping, it means that the bloods are incompatible and therefore the blood of one could not be used in a transfusion for the other. However, occasionally blood of the same basic types may be incompatible since there are other factors in the blood that can cause this incompatibility.

(C) Effect of Dilution of Serum

Make a clean slide again by drawing a line down the center and marking the left side of the slide "1/4" and marking the right hand side of the slide "1/16". Place a drop of A or B blood within each circle and add to each circle the corresponding anti-serum, diluted to the indicated strength. Use a toothpick to stir each circle. The clumping effect will disappear with dilution. (D) After doing the above, answer the following: (1) Why is it not necessary to determine the blood type when preparing for a transfusion of blood plasma? (2) Why is Type O called the universal donor? (3) What is the difference between blood clumping and blood clotting?

(*Note to Teacher:* If the drawing of blood is prohibited by your Board of Education, you may use oxalated blood, which can be obtained from a local hospital.)

Demonstration: Rh Blood Typing

Since it is not as simple to get the serum for Rh typing, do the typing as a demonstration on a microprojector to show results. (1) Prepare a slide as for blood typing but maintain slide temperature at about 37°C and not above. Place a drop of anti-Rh serum at the center of the slide. Adjacent to this but not touching, place an equal amount of suspension of red blood cells. With a toothpick mix the two drops and agitate the slide for thorough mixing. (3) Clumping is macroscopic, but the microprojector will show it more vividly. Clumping should occur within two minutes. If the blood clumps in anti-Rh, the blood is Rh positive; if there is no clumping, the test blood is Rh negative.

Read directions carefully on the serum you have to see if there are any additional directions.

AUDIO-VISUAL MATERIALS

FILMS

The Blood (J)
Endocrine Glands (J)
Principles of Endocrine Activity (GG)

FILMSTRIPS

How Hormones Control the Body (HH)
The Work of Blood (HH)

TRANSPARENCIES

Circulatory System (A)
The Endocrine System (A)

MODEL

Endocrine System (A)

KITS

Blood Typing Kit (F), (CC)
Blood Smear Kit (F), (CC)
Rh Blood Typing Kit (D), (F), (CC)

Photosynthesis

UNIT 15

No doubt you remember the first law of thermodynamics which states that matter can be neither created nor destroyed. Therefore it follows that there could be no creation of new carbon atoms to form new organized structures in existing organisms or in the structural components of new individuals. However, you must realize that organized carbon structures need be formed, or all the carbon present in the universe would be converted to carbon dioxide by metabolic processes. Thus there must be some means by which the waste products of respiration are converted to organized carbon molecules if life is to continue.

Organic structures are formed from carbon dioxide and water by autotrophic organisms, capable of producing the organic molecules necessary for their structure and function from both the basic compounds. There are several enzyme systems in different organisms that are capable of accomplishing this change with varying degrees of efficiency. The chemosynthetic organisms use the energy from chemical reactions to cause the fixation of carbon dioxide by increasing its carbon chain. Photosynthetic organisms use the energy of light to drive the carbon dioxide fixation.

It is readily seen that the most efficient means of CO_2 fixation is the enzyme system present in the photosynthetic organisms, since the process of photosynthesis is the only building process known thus far that is a highly efficient exergonic process. Therefore it is not surprising that the autotrophic organisms above the bacterial level are all photosynthetic in their fixation of CO_2.

Except for a few species of unusual bacteria which are able to capture energy from certain chemical reactions, autotrophic plants manufacture foods by virtue of their possession of chlorophyll. Photosynthesis is a process by which green plants change the energy of sunlight into the potential energy of reduced carbon compounds with the evolution of oxygen.

Chlorophyll makes possible the utilization of solar energy in combining carbon dioxide and water to form glucose or other carbohydrates, with another product in the process being oxygen. Although there are many separate chemical reactions involved, the overall reaction reads simply as follows:

$$6CO_2 + 6H_2O + \text{solar energy} \xrightarrow{\text{chlorophyll}} C_6H_{12}O_6 + 6O_2$$

This process is undoubtedly the most important process on earth, for it directly or indirectly supplies the essential nutrients for most forms of life. Because green plants are able to store the energy of sunlight, fuel substances are provided with a food source. It also supplies carbon compounds for plants and animals, and, created in prehistoric times, the vast stores of fuel such as coal, oil, and gas, which man now uses as a major source of heat, locomotion, and many other forms of power.

Although the immediate products of photosynthesis are simple carbohydrates, the green plant does not ordinarily build up great quantities of these substances. Rather they serve as raw material for the further synthesis of organic compounds. They may be converted to more complex carbohydrates or to fats in the plants, or they may be combined with nitrogen and other elements available to the plant in its environment to form proteins. Material essential to the well-being of the plant may be formed by such modification of these carbohydrates, and even more chlorophyll can be synthesized from them. It may be said that the green plant is a very able chemical factory, producing a variety of substances from such fundamental materials.

From your study of general biology you recall that the ability to carry on photosynthesis is associated with green bodies, called chloroplasts, which are suspended in the cytoplasm of the cell. If the leaf cells are ground in a colloid mill, you can see that the chloroplasts break up into smaller fragments called grana, which are green.

The grana contain, on a dry basis, about 40% proteins, 35% lipids, 7% minerals, and 5% of a green pigment called chlorophyll. It is the chlorophyll that is essential for photosynthesis. In most plants it exists as two closely related compounds, chlorophyll-a and chlorophyll-b. The formulas for both forms show that chlorophyll contains the tetrapyrole or porphyrin nucleus which is found in cytochromes, certain enzymes, and in the blood constituent, hemoglobin. In chlorophyll the coordinating ion is magnesium, whereas in the other porphyrins it is iron.

(The structural formula of chlorophyll has been shown in an earlier unit. To demonstrate the components of chlorophyll, there is a laboratory exercise which dramatically shows them by means of thin-layer chromatography.)

The chemical formula for chlorophyll-a is $C_{55}H_{72}O_5N_4Mg$. In addition, some scientists believe that there are five different kinds of chlorophyll, a, b, c, d, and e. Chlorophyll-a is thought to occur in all green plants, and these molecules are believed to initiate the process whereby the energy from the sun is used in photosynthesis. In addition to chlorophyll molecules, some plant cells contain xanthophyll and carotene pigments.

The carotenoids is the name for the other widespread family of pigments. They are named for carotene, responsible for the yellow color of corn, egg yolk, and carrots. Carotene is converted into Vitamins A, E, and K, and into visual purple. Carotenoids also cause the brilliant colors of butterflies, the red of tomatoes, and the gray-black of the ink that a squid shoots out to protect itself. More than seventy carotenoids are known, all with similar structures.

If you look at the general equation for photosynthesis, you will notice that it is the reverse of respiration. The water molecule shown on the right side of the equation contains oxygen which was originally in the CO_2, and therefore not identical with the H_2O shown on the left-hand side of the equation. Through experimentation it was proven that the oxygen released during photosynthesis comes only from water consumed in the photosynthetic reaction and does not come from the CO_2 consumed. The oxygen of the CO_2 consumed appears in the carbohydrate and in the water formed during photosynthesis.

The process of photosynthesis may be regarded as two coupled reaction series. The first reaction series requires light and involves the generation of energy from this exposure to light. It is known as the *light reaction*. The other series is dependent upon energy generated from the light reaction, but *does not require* light in order to proceed. This is called the *dark reaction*. The dark reaction involves the CO_2 fixation reactions and the ultimate formation of glucose by means of a set of reactions closely analogous to a reversal of part of the anaerobic glycolysis scheme. In the dark reaction it is a series of enzyme-catalyzed reactions. (There are a number of demonstrations to use in order to show what is needed and used in photosynthesis.)

LIGHT REACTION

Sunlight or a source of light energy is necessary for the light phase to occur. Energy from the sun reaches the earth as radiations with different wavelengths. Certain of these wavelengths constitute white or visible light (wavelengths from 390-76 mu). White light is a mixture of red, orange, yellow, green, blue, indigo, and violet waves. These colors are known as the visible or solar spectrum.

Chlorophyll is a pigment with the capacity to absorb certain of the wavelengths in visible light. Green land plants absorb mainly the red and blue-violet wavelengths, and these are the most effective ones for photosynthesis. The energy absorbed by chlorophyll is used in the photolysis of water.

Photolysis is the first step in photosynthesis and is the breakdown of water in the presence of chlorophyll, with light energy serving to activate the reaction:

$$2H_2O \xrightarrow[\text{chlorophyll}]{\text{light}} 4H + O_2$$

The light reaction involves the excitation of chlorophyll-a by light. This excitation of the chlorophyll causes a split of a molecule of water, forming hydrogen, free oxygen, and an electron. The transfer of the electrons formed from the splitting of the water is not unlike that of the electron transport system and ultimately generates energy in the form of ATP.

Chlorophyll-a is the only photosynthetic pigment capable of undergoing this energy-yielding reaction. Other photosynthetic pigments present in the chloroplast, chlorophyll-b, xanthophyll, and carotene, are excited by light and transmit this excitation to chlorophyll-a. The advantage of the presence of the other pigments is their ability to absorb different wavelengths or colors of light:

$$\text{chlorophyll-a} \xrightarrow[\substack{\text{or excited auxiliary} \\ \text{pigments}}]{\text{light}} \text{"excited" chlorophyll-a} + \text{generation of ATP}$$

The oxygen produced in this reaction may escape into the immediate environment of the plant, or some of it may be used in the plant for other reactions. As for the hydrogen, it is captured by certain molecules that serve as hydrogen acceptors and eventually delivered to a complex cycle of reactions, which CO_2 also enters.

In this primary phase of photosynthesis, the radiant energy trapped in chlorophyll serves to synthesize two compounds needed for the actual incorporation of CO_2 into carbohydrates. The first of these, ATP, is produced by the photophosphorylation of ADP.

The second compound is the reduced form of nicotinamide adenine dinucleotide phosphate (NADP) which arises by a photochemical reduction of the oxidized (NADP +) form. Both these steps consume energy and can only take place if this energy loss is compensated for by some form of energy supply, such as the energy of light. (The Hill reaction illustrates perfectly this phase.)

NADP is a coenzyme and it cannot be formed without the aid of nicotinic acid; without NADP many metabolic pathways could not be maintained. NADP is also known as triphosphopyridine nucleotide (TPN).

DARK REACTION

The dark reaction has a name that is misleading since the reaction will take place equally well in light or darkness. It is an enzyme-controlled reaction involving the reduction of carbon dioxide to carbohydrate. Carbon dioxide is taken up by a three-carbon compound already present in the chloroplast, called ribulose diphosphate (RDP). A molecule of RDP, one of CO_2, and one of water react to produce two three-carbon molecules of a compound called phosphoglyceric acid (PGA). Each PGA molecule loses an oxygen atom (which is joined to hydrogen coming from photolysis) and becomes phosphoglyceraldehyde (PGAL). By special transformation reactions, five out of six PGAL molecules produced in this fashion are changed to three of RDP, which then go on for more CO_2. One out of six is made available to the plant.

Actually PGAL is the end product of photosynthesis, not glucose as the overall equation would indicate. However, since PGAL may be converted to glucose, it is accurate enough for the purposes of representation to balance the equation as we do. To summarize, the molecules produced by the reaction are not the identical molecules that entered it. The sugar and water molecules on the right side of the equation are completely new units. What happened primarily was that an increase of energy is brought about in the system in order to incorporate the carbon dioxide. This energy is available as the food supply of the photosynthetic organism. It is also available to whatever heterotrophic organisms are able to grow upon it. In the dark phase, the energy is used to form a more complex organic product.

The requirements for photosynthesis are: (1) carbon dioxide as a source of carbon; (2) light as the energy source; (3) chlorophyll as the site of energy absorption; and (4) water as the source of the electron necessary for the generation of energy in a usable form.

The essential features of photosynthesis are: (1) energy in the form of activated electrons, which are used to (2) incorporate carbon into an already existing compound that will eventually become glucose or starch; (3) oxygen, which is a by-product of water breakdown; and (4) water, another by-product formed by the union of the hydrogen from the required water and oxygen from the carbon dioxide.

There are several factors which may limit the rate of photosynthesis: (1) Under low light intensity, light becomes a limiting factor. (2) Temperature affects the dark reaction, which involves the action of many enzymes. It is a limiting factor for all plant growth in the early spring and late fall in most parts of North America. (3) Water is an essential reactant; if the leaves wilt, the rate of photosynthesis will decline. (4) Carbon dioxide reduction will decrease the rate of photosynthesis. (5) Chlorophyll is essential for photosynthesis. In variegated leaves, experimentation shows that no photosynthesis takes place. (6) Carotenoids are not essential for the light reaction of photosynthesis; however, light energy absorbed by the carotenoids can be transferred in part to the chlorophyll. (7) Enzymes are also necessary for the process of photosynthesis.

(Many of these factors can easily be demonstrated to the class. There also are a number of interesting laboratory exercises that the class can do to better study the various components of chlorophyll and other pigments found in plants, all of which have some part in photosynthesis.)

Demonstration: Carbon Dioxide and Photosynthesis, Using Brom Thymol Blue

The pH range of brom thymol blue is between 6.0 and 7.6. At 7.6 it is blue, and turns yellow as it approaches 6.0. The carbon dioxide in water is carbonic acid and the reaction with brom thymol blue may be summarized: brom thymol blue + carbonic acid becomes brom thymol yellow.

To show that plants use the carbon dioxide dissolved in water as carbonic acid, blow into a test tube of brom thymol blue until the solution turns yellow. As soon as you note the yellow color, stop. Place a sprig of actively growing Elodea into the test tube of brom thymol yellow, and place the test tube in front of a 250-watt bulb in a reflector hood or place the tube in the sunlight. Within an hour the brom thymol yellow should turn blue as the plant uses the CO_2. As a control, use another tube with brom thymol yellow and no plant. Nothing happens.

Demonstration: Light and Photosynthesis

Set up two tubes with brom thymol yellow and into each place a sprig of Elodea. Then cover one tube completely with black paper so that no light can penetrate and place the other either in the sun or in front of a strong light. After an hour the one without the paper will turn blue as the plant uses the CO_2 with the aid of light, while the other tube with no light will still be yellow, showing that the plant did not carry on photosynthesis and use the carbon dioxide.

Demonstration: The Need for Light

In order to produce glucose, light is needed. In this demonstration the starch is tested as an indirect evidence of photosynthesis.

Cut two pieces of aluminum foil slightly larger than a geranium leaf. Place one above and one below the leaf and crimp the side so that there is no light hitting the leaf. Keep the plant in the sunlight for several days. Cut the covered leaf from the plant, remove the foil, and cut the leaf into several pieces. Place the leaf pieces in a beaker with 50 ml of acetone to extract any coloring in the leaf. Test the leaf strips with Lugol's solution. If there was starch present the leaf would turn blue-black, which it did not.

If you were to use the Lugol's solution as a test for starch on a leaf that has been left uncovered, it would turn blue-black, showing starch was present. Thus without light no starch could be manufactured.

Demonstration: Chlorophyll Is Necessary for Photosynthesis

Use either a leaf from a variegated Coleus plant or a silver leaf geranium plant that has been in strong light for several hours. Extract the pigments by placing the leaf in a beaker of hot alcohol over a hot plate in a beaker of water. If the Coleus is used, remove red pigments first by placing the leaf in hot water and transferring to the beaker of alcohol.

After the pigments have been extracted, place the leaf in a Petri dish and pour iodine over the leaf. Where there was green in the leaf there will be a positive test for starch, and where there was no green there will be no reaction with the iodine.

Demonstration: Oxygen Evolved in Photosynthesis

Invert a three inch tip of vigorously growing Elodea sprig into a test tube containing aquarium water, to which about 2 ml of 0.25% solution of sodium bicarbonate has been added for 100 ml of aquarium water. The water should first have been boiled to drive off dissolved gases and then cooled. The bicarbonate will be a source of carbon dioxide. Tie the sprig to a glass rod immersed in the test tube so that it will be held down. Expose the plant to a bright light. Bubbles of gas will be seen escaping from the cut stem; count how many per minute.

Then remove the test tube further from the light and count the bubbles. Keep moving the test tube and you will see that the bubbles are less and less as the test tube is moved further and further from a light source.

Demonstration or Laboratory Exercise: The Hill Reaction

Aim: To study the role of light reaction in photosynthesis as demonstrated by Robin Hill.

Introduction

The first notable contribution in the study of photosynthesis was made in 1905 by the English plant physiologist, Blackman. He demonstrated that photosynthesis is not a single photochemical reaction. It consists of two major reactions: the first is a rapid reaction and requires light energy for its acceleration, while the second reaction does not use light energy and goes on equally well in either light or dark. However, it was not until 1937 that Robin Hill, an English biochemist, showed the possible nature of the light reaction.

Illumination of suspensions of chloroplasts in water in the presence of a suitable hydrogen acceptor (oxidant) results in the release of oxygen. This phenomenon, now generally called the Hill reaction, occurs also in suspensions of grana from disintegrated chloroplasts. Some of the compounds which act as hydrogen acceptors are ferrocyanides, chromates, certain quinones, and certain indophenols. By use of the 0-18 isotopes of oxygen, it has been shown that the oxygen released in the Hill reaction, as in the overall process of photosynthesis, comes from the water molecules.

The Hill reaction apparently consists of a photocatalyzed splitting of water molecules. A probable role of this reaction in the overall process is the formation of hydrogen atoms, which are stored in the chloroplasts by combinations with some hydrogen acceptors. Subsequent chemical transformation involving hydrogen transfer utilizes the hydrogen which has been formed and captured in this manner. Present evidence indicates that the hydrogen acceptor is triphosphopyridine nucleotide, TPN.

The primary role of light is the photolysis of water. The period during which CO_2 reduction can occur in the dark at the expense of a previous exposure to light is of very short duration.

Because the Hill reaction indicates that fragments of chloroplasts in the light decompose water to hydrogen and oxygen, it is of great interest to biology students.

The procedure is to permit a suspension of chloroplasts from spinach to react in the light with the blue solution of indophenol (2, 6, dichlorophenol indophenol). As the indophenol takes up hydrogen, it becomes reduced to a colorless form. By implication oxygen is also released, but no test for the oxygen is made in this procedure, nor are the bubbles of oxygen observed. The controls include the use of indophenol without chloroplasts and an additional pair of tubes in the dark.

Procedure: Since this is a rather complicated procedure, the following is for one demonstration, but several teams can do the exercise if desired.

(1) Prepare a solution of 0.01% indophenol by dissolving 0.1 gram indophenol in 1000 ml of demineralized water. Place this solution in a stoppered dark bottle and refrigerate until needed. (2) Prepare a cold solution of 10% sucrose by dissolving 10 grams of sucrose in 50 ml of demineralized water and adding ice cubes to the 100 ml mark. (3) Place 25 grams of fresh spinach in a blender and add 100 ml of the cold sucrose solution (including ice cubes). Homogenize for thirty seconds at high speed. (4) Filter the homogenate through two layers of cheesecloth placed over a beaker. (5) Centrifuge the filtrate for five minutes at high speed. Meanwhile, prepare another 100 ml of cold sucrose as described in step 2. (6) While step 5 is in progress, prepare four test tubes of cold indophenol solution, as follows: to 10 ml of ice water in each tube add 2 ml of the previously prepared indophenol solution. (The indophenol could be diluted to start with, but if this is being done

as a laboratory exercise, it gives the student practice in using pipettes and making dilutions.) (7) After the chloroplast suspension is centrifuged, pour off the supernatant fluid and resuspend the chloroplasts in 2 ml of cold sucrose solution. This is now the relatively cell-free chloroplast solution which is used in the main portion of the exercise. (8) Now mark the tubes as shown in Figure 15-1. (9) Add the cold chloroplast suspension to tubes 2 and 4; using a medicine dropper, add only one or two drops of the suspension prepared in step 7 and thoroughly mix the contents of each tube. A slight green tinge develops in these two tubes because of the chloroplasts. Wrap the control tubes 3 and 4 completely in aluminum foil; their contents are to receive no light. (10) Exposure to light: Place all four tubes into a beaker containing ice water and floating ice cubes. Expose to a 150 watt incandescent bulb placed a few inches from the beaker. Within ten minutes tube 2 is decidedly more pale in color than tube 1. If the change is not apparent, add one additional drop of the chloroplast suspension to tube 2 and shake.

Results: Chloroplasts in the light produced a substance which reduced indophenol since tube 2, containing the chloroplasts and exposed to the light, is a pale green. Tube 1, without the chloroplasts, is the same intensity of blue as tube 3. Tube 2 is much lighter in intensity than tube 4.

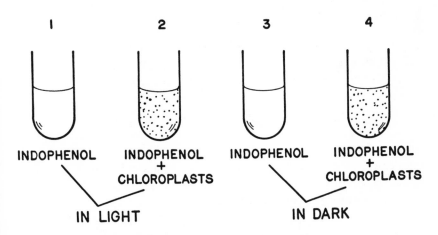

Figure 15-1 The Hill reaction

*Laboratory Exercise: Chromatography and
Fluorescence of Plant Pigments*

Aim: (1) To separate several of the pigments of a plant, and (2) to observe the phenomenon of fluorescence of chlorophyll.

Introduction

Pigments   Among the pigments present are the following: *chlorophylls,* which include chlorophyll-a, chlorophyll-b, chlorophyll-c, chlorophyll-d. These are various shades of green. *Carotenes,* which may be white, yellow, orange, or red. *Xanthophylls,* which are usually bright yellow. *Anthocyanins,* which are water-soluble pigments present free in the cell cytoplasm instead of plastids and may be red, blue, or purple. In the autumn, when the chlorophyll in leaves disintegrates, the colors of the xanthophylls and the anthocyanins become unmasked, causing the autumnal display.

Fluorescence   Figure 15-2 reviews your understanding that white light is a mixture of various wavelengths. These may be dispersed by a prism into the component wavelengths to form the visible spectrum. An object which looks green is one which absorbs all other wavelengths and *reflects* (or transmits) the wavelength which affects our eyes and brain with the sensation of green. When chlorophyll reflects green wavelengths, it is absorbing the other wavelengths of red, orange, yellow, blue, indigo, and violet. Since light is a form of energy, the wavelengths that are absorbed have the capacity to do work. Plants exposed only to green light cannot carry on photosynthesis, and soon die.

The structure of the chlorophyll-a molecule is such that it can capture light energy. This seems to be by some form of resonance of the atoms comprising the molecules. Examine the structure of the chlorophyll molecule in Figure 15-3. Observe that it is composed of four complex carbon-hydrogen rings which are joined into a large ring, the "head." In the center of the head is a magnesium atom. It is interesting and probably of evolutionary significance to point out that there is another kind of molecule which greatly resembles the chlorophyll head, except that there is

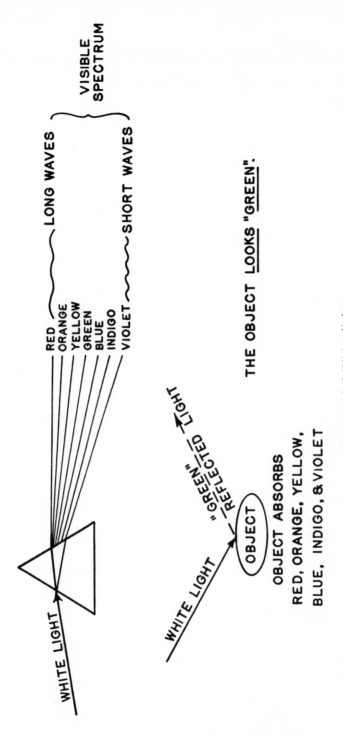

Figure 15-2 White light

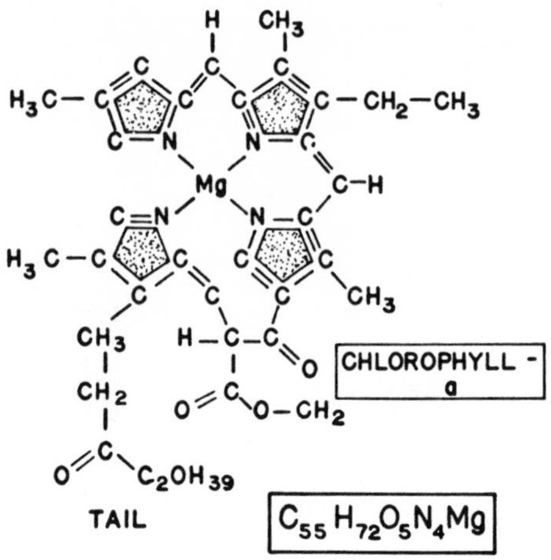

Figure 15-3 Chlorophyll molecule

an iron atom in place of magnesium. This is the *heme* molecule, the colored component of hemoglobin. The remainder of the chlorophyll molecule is a long "tail" of $C_{20}H_{39}$.

When red-orange and blue-violet light is absorbed by the atom of the chlorophyll molecule, the atom may become *excited;* that is, an electron may be displaced from its normal orbit farther away from the nucleus of the atom (Figure 15-4). This excited state lasts about 10^{-10} seconds. When the electron reverts from the excited state to its position, the atom releases the energy which it has absorbed. Most of it presumably is used in splitting water (Figure 15-5), but about one percent is emitted as a brief flash of light called *fluorescence.* Inasmuch as the light which is emitted has less energy than the light which was originally absorbed, it has a longer wavelength and thus appears red.

Preparing the Chloroplast Pigments

Materials Needed: Dried parsley leaves, 10 grams; 25 ml acetone; 50 ml diethyl ether; mortar and pestle; funnel; beaker. The dried parsley leaves may be obtained from the spice shelf of any supermarket. The brand is not important; all give the same excellent results.

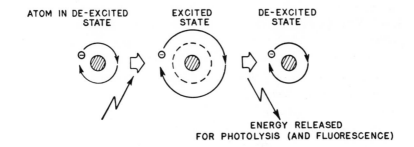

Figure 15-4 Excitation of atom

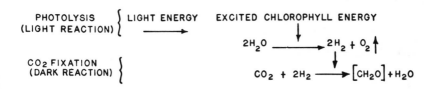

Figure 15-5 Splitting water

Procedure: (1) Weigh 10 grams of parsley and pulverize it to a fine powder with a mortar and pestle. Place the leaves in a beaker. (2) Add 25 ml acetone and 50 ml diethyl ether (sulfuric ether) to the powdered leaves. Mix thoroughly. (3) Allow the mixture to stand, covered, for at least twenty minutes, then mix again. (4) Filter the mixture. The filtrate is the chlorophyll extract that will be used for chromatography. (5) Store the extract in brown cork-stoppered bottles (rubber stoppers deteriorate) in the refrigerator. Colors are more vivid when the extract is left undisturbed for several days before use.

Preparing the Microscopic Slides (Microchromatic Plates)

Materials needed and procedure are to be found in the first unit of this book, and there will be a quantity sufficient for a class of twenty.

Spotting the Slides   Only minute amounts of the chloroplast solution are needed for the chromatoplate. Satisfactory
spotting pipettes for school use may be made by drawing out
melting-point tubes with the aid of a Bunsen burner.

(1) Fill the pipette with the chloroplast solution by capillary
action. The first drop may be too large and should be extruded
onto a piece of scrap paper. (2) In placing a drop on the silica gel,
try to touch only the liquid on the surface without breaking the
surface. Ideally, each drop should be small enough to spread no
more than 1 to 2 mm in diameter. Since the chloroplast pigments
are dissolved in volatile acetone-ether, the drop dries immediately
without the need for hot-air driers. (3) Arrange twenty drops
adjacent to each other to form a line across the plate. The spotting
should be about 1 cm from the bottom of the slide. (*Caution:* The
room should be well ventilated at all times because of the extreme
volatility of the liquids used.)

Developing the Chromatogram

Materials Needed: Two volumes iso-octane, one volume acetone,
one volume diethyl ether, Coplin jars, micropipettes, coated slides.
(*Note:* Twenty-five milliliters of this developing solvent should be
sufficient for the Coplin jars for the entire class. *Caution:* Have no
flames in the room.)

Coplin jars are convenient to use as developing containers,
though the grooves should not be used since they will break the
silica gel film. If these are unavailable, any small covered jars may
be used. Aluminum foil or Saran Wrap may be fashioned into
covers.

Procedure: (1) Into the chromatojar, introduce the developing
solution to a depth of about 0.3 cm. For a Coplin jar, this requires
a volume of about 2 ml. The height of the liquid in the jar should
be below the spots on the microscope slide. (2) Cover the jar for a
few minutes so that the vapors fill the container. (3) Now insert
the spotted microscope slide into the chromatojar and cover the
jar. (4) The solvent rises, and within two minutes a brilliant array
of green and yellow bands appear. Within five minutes the solvent
front approaches the top of the silica gel layer and the slide should

be removed for examination. *Note:* The chromatogram is air-dried within one minute. No ovens or developing agents are needed. The pale yellow bands may begin to fade in light, but good observations can be made for several hours. Some slides are good for several days.

Analyses of Chloroplast Pigments

Identification of the Bands   For the average high school class, it is unnecessary to go beyond the observation that green chloroplast solution has been separated into blue-green chlorophyll-a, green chlorophyll-b, yellow carotenes, and yellow xanthophylls.

Rate of Flow   The rate of flow is the ratio of the distance traveled by the dissolved substances to the distance traveled by the solvent. To measure the Rf of the various components of chlorophyll, measure the distance the solute traveled from the original dot. Each band will have traveled a different distance. Then measure the distance the solvent traveled from the spots.

The Rf of any substance is constant for a given set of circumstances. When all of the distances have been measured, the Rf of each substance may be calculated individually by using the following formula:

$$\text{Rate of flow (Rf)} = \frac{\text{Distance traveled by solute}}{\text{Distance traveled by solvent}}$$

Absorption of Light by Chlorophyll   The mixture of chloroplast pigments may be used for roughly indicating the wavelengths which are absorbed by chlorophyll. Shine a strong white light (as from a slide projector) through a triangular prism onto a screen to form a spectrum. Now interrupt the white light with a bottle of chloroplast pigments. The green portion of the spectrum becomes much more dim, indicating the absorption of the wavelength for green.

To demonstrate fluorescence, use the same mixture of chloroplast pigments. When exposed to strong white light, the green fluid appears red when seen from certain directions. (The mixture should be placed in a square-sided bottle.)

AUDIO-VISUAL MATERIALS

FILMS

How Green Plants Make and Use Food (K)
Photosynthesis (B), (J)
Photosynthesis: The Biochemical Process (K)
The Riddle of Photosynthesis (B), (U)

FILMSTRIP

Photosynthesis (The Living Cell Set) (A)

SINGLE-CONCEPT FILM LOOPS

Carbon Dioxide Requirement: Photosynthesis (J), (AA)
Chlorophyll Requirement (J)
Factors Affecting Oxygen Production (J)
Light Requirement for Starch Production in Green Plants (J), (AA)
Photosynthesis Fixation of CO2 (A)
Photosynthesis Set (A)

TRANSPARENCIES

Dark Reaction (F)
Light Reaction (F)

MODEL

Chloroplast (CC)

KITS

Basic Chromatography of Pigments (EE)
Chromatography I and II (A)
Chromatography and Electrophoresis Kit (U)
Chromatography Kit (F), (S)
Thin-Layer Chromatography (S)

Plant Digestion and Plant Hormones

UNIT 16

At this point the class should have realized that since plants do not obtain food carbohydrates by a process of extracellular digestion such as that occurring in micoorganisms and animals, the carbohydrates must be supplied by photosynthesis in the leaf. The carbohydrates are glucose, fructose, and sucrose, which are transported to the plant parts requiring these nutrients. If some of the carbohydrates formed during photosynthesis are not needed to build structural material like cellulose or protoplasmic materials like proteins and lipids, they are stored as polysaccharides, which are usually starch. However, the excess can be stored as sucrose if the plant does not form starch in the cells of a particular tissue.

CARBOHYDRATES

This digestion in plant tissues is concerned with intracellular digestion of stored food carbohydrates. The most important sites of starch storage are in seeds, tubers, and in the leaves during photosynthesis. The digestion of stored starch is most rapid during the germination of seeds and sprouting of tubers. The carbohydrate stored in leaves is used when the plant is placed in the dark and cannot carry on photosynthesis.

There are two ways in which starch is broken down to glucose. In pea seeds and potato tubers it has been found that the enzyme *phosphorylase* splits the starch molecule by introducing phosphate groups at the glycosidic linkages. Since the reaction is reversible under certain conditions, the phosphorylase enzyme can synthesize an amylose type of starch. Another enzyme is present to form the amylopectin type from the amylose type.

In the second method, the enzyme *beta-amylase* attacks the nonreducing ends of the starch molecules; the successive units of maltose are split off until a point is reached at which amylose is completely split to maltose, and, in amylopectin, until a branching of the chain occurs. The maltose, with the aid of maltase, then becomes glucose.

In those plant parts where sucrose serves as the storage form of carbohydrates, the enzyme *sucrase* (invertase) is probably involved in splitting the sucrose molecule into glucose and fructose. Invertase is widely distributed in plant tissues. Sucrose is readily used for growth in intact plants, excised roots, various plant tissues, and young embryos, and is phosphorylated with ease.

In higher plants carbohydrate breakdown is similar to that found in animals. Higher plants contain enzymes which are able to convert glucose to ethyl alcohol and carbon dioxide, and this conversion takes place when plants are in an oxygen-free environment. The plant might live for a limited period of time.

With the exception of starch, glucose, fructose, and sucrose, little is known regarding the mechanism of synthesis of the various carbohydrates found in plants. All of the carbon compounds of a plant can be synthesized, starting with glucose, fructose, or sucrose. (Since this entire topic can be discussed in class in a short period of time, do the demonstrations individually and the laboratory exercises when the unit is finished without being concerned that the students might not completely remember the basic concepts learned. Plant enzymes are in this unit.)

LIPIDS

Plants do not have an outside source of lipids; they must synthesize them as needed. Only in the germinating seed does digestion of stored lipids take place. It has been found that during germination, the lipases found in the seed split stored triglycerides to diglycerides and monoglycerides, free fatty acids and glycerol, and the carbon of these compounds is converted to carbohydrates, amino acids, new lipids, and other chemical building blocks as needed by the developing seeds for its vegetative growth. The lipids are widely distributed in plants, but in general seeds contain more crude fat than other plant parts. Most of this fat is true fat, triglyceride. In general a seed will store a large amount of either true fat or starch, but not both. There are some wax compound lipids and derived lipids to be found in certain leaves and stems.

PROTEINS

The major uses of protein in lower plant life are as nucleoproteins in the genes and as enzymes in the various parts of the cell such as nucleus, cytoplasm, mitochondria, ribosomes, etc. In the higher plant forms proteins have an additional use as a reserve of amino acids in the seeds.

Plants synthesize all the amino acids necessary from carbohydrates and a simple nitrogen source such as ammonium ion, nitrate ion, etc. The legumes are even less dependent, since they are supplied with fixed nitrogen by symbiotic bacteria living in the root nodules and fixing atmospheric nitrogen.

It is believed that all forms of fixed nitrogen are converted to ammonium ion, either free or combined, before being incorporated into amino acids. The reduction of nitrate to ammonia may follow this pattern: $NO_3 \rightarrow NO_2 \rightarrow NH_4OH \rightarrow NH_3$.

Through experimentation with legumes there is evidence that the legume obtains most of its nitrogen from the root nodule bacteria in the form of aspartic acid. The plant tissues contain enzymes, called transaminases, which in turn form alanine. However, the transamination reaction does not take place with all of the amino acids, as certain ones such as arginine and tryptophan are synthesized by reactions involving direct amination.

Thus nitrogen is supplied to the plant from three sources, but regardless of the source, all forms of nitrogen are converted to NH_3 before reaction with the appropriate alpha-keo acid to form an amino acid. The alpha-keo acids, which serve as precursors of the amino acids, are primarily products of carbohydrate metabolism, but also can come from the deamination of other amino acids or from fatty acids, especially in germinating seeds.

Once the amino acids are synthesized, they are linked together to form the proteins needed by the growing plant. Protein formation requires energy and probably proceeds via a series of phosphorylation reactions.

AUXINS

Auxins are plant growth hormones. The three most common ones are 3-indoleacetic acid (IAA), 3-indoleacetonitrile, and traumatic acid. Although other hormone substances have been reported to be in plants, their presence has not been verified. Plant hormones are believed to cause the following responses in plants: phototrophism (bending toward light); geotropism (bending of stems away from, and roots toward, the center of gravity); bending produced by injury (medicated by traumatic acid); spinasty (downward bending of leaves); initiation of flowers; inhibiting of sprouting; and the hastening of rooting.

The auxins are produced in the tips of roots and stems, and in rapidly enlarging fruit. The auxins may either stimulate or slow down growth.

IAA stimulates growth of the stem tip, but at the same time prohibits development of the buds. IAA is produced in the growing tip of the stem, although it may occur in lesser concentrations throughout the plant. It is continuously produced in the growing plant and translocated from the site of production to other parts of the plant as needed. However, the enzyme *IAA oxidase* prevents a high concentration of the enzyme from accumulating. Thus the oxidase is an indirect regulator of stem growth. This inhibition of concentration of IAA is important, as a trace of the hormone applied to one side of a stem greatly accelerates growth on the treated side.

IAA has a growth effect on young cells only as the presence of heavy cellulose deposits in older cells prevents this effect. It is believed that the hormone enables a young cell to maintain the plasticity of its primary cell wall and thus prevent the cell from elongating through the accumulation of cellulose. As long as equal amounts of cellular IAA are present in all the cells at a given level of the stem tip, there would be uniform elongation on all sides and a straight stem would result. However, if unequal amounts of cellular IAA are present at any given level, unequal growth would occur at that place on the stem and the stem would follow a curved path of growth.

Light energy accelerates the action of the IAA oxidase so that the sides of the stem upon which light falls will have a lower concentration of cellular IAA than will the shaded side. The higher level of auxin on the shaded side will promote greater growth on the shaded cells, and the stem tip will curve toward the light until all sides of the stem are equally illuminated. When the IAA oxidase is equally stimulated by the light, the auxin level on all sides is equalized and the stem grows straight toward the light source.

Auxin molecules, as all other molecules, respond to gravity and are pulled downward. Their concentration will increase on the lower side of the stem when a plant is placed on its side, and growth will be accelerated on the lower side to produce an upward curvature of the stem.

IAA also effects root elongation but high levels of auxin inhibit root cell elongation. Auxin also inhibits the growth of the lateral buds. This is evident if you were to cut off the growing tip of some garden plant. The removal of this growing tip will free the lateral buds from dormancy which is imposed by the presence of the stem tip, and side branches will soon develop. This is the reason for the pruning of trees, bushes, shrubs, and flowers in order to make them grow thicker.

Recently scientists discovered another plant hormone, known as cytokinins. Some of the cytokinins chemically resemble adenine, an essential component of DNA. Although there is a more pronounced effect when cytokinins, auxins, and gibberellins are used in various combinations, it has been found that the cytokinins specifically promote cell division and control differentiation of the meristematic cells.

Gibberellins

There are synthetic auxins which can alter the growth of useful plants. Very popular growth substances used by botanists are the gibberellins. Gibberellins are produced by a fungus, *Gibberella fujikuroi*. At least three gibberellins have been isolated from this fungus and all are acids with complex structures.

A great deal of work has been done with gibberellins, and it has been found that the gibberellins are able to produce a marked elongation of the plant's stems. If a dilute solution of gibberellic acid is applied to five pea seedlings of equal length prior to planting them in a pot, while another five are used as a control in another pot, the seedlings treated with the gibberellic acid will have considerably more growth.

Another growth hormone, although synthetic, has been found to be a herbicide, 2,4,-dichlorotophenoxyacetic acid, commonly known as "2,4-D." Broad-leafed plants of the dicots are much more sensitive to this growth hormone than the monocots, such as grasses and cereals. Thus when 2,4-D is applied to lawns and pastures it acts as a weed control for the broad-leafed weeds. There is also a compound effective for the killing of resistant woody, noneconomic plants of the brush and shrub types. This is 3,4,5-trichlorophenoxyacetic acid (2,3,5-T).

Demonstration: Phosphorylase and Starch Synthesis

The enzyme *phosphorylase* is found in plants and animals and catalyzes the change of glucose-1-phosphate to starch and inorganic phosphate. (1) Into two Petri dishes pour 15 ml of 1% glucose-1-phosphate plus several ml of toluol to prevent bacterial contamination. Also into two Petri dishes place 15 ml 1% glucose. (2) Now soak some soybeans for twelve hours. Test several soybeans for starch by cutting thin sections from the bean and testing the sections. (3) If no starch is found, place the remaining sections into the Petri dishes with glucose-1-phosphate. Also prepare the soybeans in a similar manner for the Petri dishes with the glucose. (4) Incubate all the dishes at 25°C. Withdraw several sections from each dish at five-minute intervals for one hour. Test the sections with IKI for starch.

The appearance of a blue color will indicate the presence of starch granules. There should be no activity in the sections from the glucose media dishes; glucose must be phosphorylated before it can be converted to starch by starch phosphorylase.

Demonstration: Diastase, a Plant Enzyme

When a seed starts to sprout, the cotyledon secretes an

enzyme called diastase which acts on the starch to change it into sugar, which can be absorbed by the baby plant as it grows.

(1) To prove this, sprout several corn seeds or some other grain. (2) After several days grind the sprouted seeds in a mortar. Empty the contents of the mortar in a small beaker or glass and add enough water to cover. Let this stand for about an hour and then filter. Save the clear liquid (filtrate). (3) To test for the presence of diastase, prepare a dilute starch paste by dissolving about 1/8 teaspoon of starch in a cup of water and boil it until clear. Cool. When cool, fill three test tubes about 1/4 full of starch. (4) Test the first test tube for the presence of starch by adding a few drops of iodine solution. It is positive. (5) Now to the second and third test tubes add 4 ml of the enzyme solution and let stand for about thirty minutes in a warm place. Check the second tube for presence of starch with iodine and it should be almost negative. Check the third test tube for sugar using Benedict's solution; it should be positive, showing the change.

Demonstration: Sucrase, a Plant Enzyme

The yeast plant will furnish the enzyme, sucrase, which acts to change sucrose. A similar enzyme, invertase, is produced in your body by the intestinal glands. To obtain sucrase, put a small piece of yeast cake or a few grains of granular yeast in a cup of water. Let stand for twenty minutes, stirring occasionally. Filter and save the clear filtrate (the clear liquid) which contains the sucrase.

To check if sucrase is present, make a sugar solution by putting 1/8 teaspoon of cane sugar in a cup of water. Fill two test tubes about 1/4 full of sugar solution. Test the contents of the first tube with Benedict's solution for the presence of simple sugar. It is negative. To the second tube add about 6 ml of the enzyme extract solution and allow to stand for an hour. Test again with Benedict's; it will be positive, showing that the cane sugar was changed.

Demonstration: Amylase

Germinate barley seeds and allow to grow for five days. Then take fifteen seedlings and grind them in 50 ml of distilled water and clean sand by using a mortar and pestle. Pour off the amylase extract fluid and centrifuge for five minutes at 500 x g. Then test a series of dilution of the amylase on 0.4% soluble starch solution. Every thirty seconds, test a sample of the amylase-starch mixture for the disappearance of starch by means of Lugol's solution for starch or Benedict's solution for the appearance of sugar. You will find the starch disappearing and sugar appearing in its place.

Demonstration: Hydrolysis of Starch in Plants

(1) Take ten corn grains, soak, and then germinate for about ten days prior to demonstration. (2) Take another ten corn grains and soak them for two days prior to demonstration. (3) Take ten dry corn grains and grind the grains in a mortar with fine sand in 10 ml of water. Filter and place half the filtrate in a test tube marked #1, the other half in a tube marked #2. (4) Take the corn grains soaked two days and grind them with fine sand in 10 ml of water. Place half the filtrate in a test tube marked #3 and the other half in a tube marked #4. (5) Finally, take the ten-day soaked corn grains and grind them with fine sand in 10 ml of water. Place half the filtrate in a test tube marked #5 and the other half in a tube marked #6. (6) Test numbers 1, 3, and 5 for the presence of sugar with Benedict's solution. #1 will definitely be negative, #3 will show some trace of sugar, and #5 will definitely be positive for sugar. (7) Test numbers 2, 4, and 6 for starch with Lugol's solution and #2 will be very positive, #4 partially positive, and #6 will be definitely negative, showing the change in the starch as the seed germinates.

Demonstration: Protein-Splitting Enzyme in Plants

Mash some slices of fresh pineapple and collect the juice, which will contain bromelin, a protein-splitting enzyme. Add equal amounts of the juice to two test tubes containing boiled egg white. At the same time boil a similar volume of pineapple and add this boiled juice to chopped egg white in two other test tubes. Add two to four drops of toluene to all four tubes to retard spoilage. Within a few days there will be partial digestion of the egg white in the test tubes containing fresh pineapple juice.

Laboratory Exercise: How Plant Cells Make Starch From Sugar

Aim: To learn whether starch found in plants has been synthesized from the sugar produced in the process of photosynthesis, or if this synthesis is independent of photosynthesis.

Materials Needed: Paper carton, black construction paper, albino tobacco seeds, iodine solution, 8% glucose solution (by weight), flats for germinating seeds, test tube racks, Petri dishes, agar, glucose-1-phosphate, cheesecloth, sawdust, electric blender, potato slices, beakers, funnel, filter paper, razor blades.

Procedure: (1) Start the seeds germinating in sawdust. (2) Prepare your dark box by either lining a large paper carton with black construction paper or covering the floor with plastic and draping flaps of paper over the open side, which will fold over to cover the contents of the box. The box should be placed near a source of heat. (3) Prepare a medium containing 2% agar and 0.5% glucose-1-phosphate in water. Boil the medium, pour into Petri dishes, and allow to solidify. (4) Prepare the potato extract by putting potato slices into the blender and blending to a pulp. Strain through several thicknesses of cheesecloth into a beaker. Allow the beaker to stand for several hours so that the starch grains remaining will settle to the bottom of the beaker. Decant the filtrate carefully into another beaker so that the juice is starch-free. Test a few drops with iodine to be sure this is so. (5) When the seedlings have reached a height of two inches or so, remove them from the sawdust and remove any of the kernels still remaining in order to eliminate any source of sugar. (6) Number the test tubes from one to eight and treat each one as follows: #1—green seedling in water and kept in the light. #2—albino seedling in water and kept in the light. #3—green seedling in water and kept in dark box. #4—albino seedling in water and kept in the dark box. #5—green seedling in glucose solution and kept in the light. #6—albino seedling in glucose solution and kept in the light. #7—green seedling in sugar solution and kept in the dark box. #8—albino seedling in sugar solution and kept in the dark. (7) Keep the dark box and the test tube rack side-by-side in a warm place for twenty-four hours. (8) Meanwhile, to demonstrate enzyme action, place four drops of the potato extract, separately and well spaced, on the agar medium in the Petri dish. The extract contains phosphorylase, which is an enzyme that converts sugar to starch. (9) Test one drop immediately for starch, using the iodine solution. Test the remaining drops one at a time at fifteen-minute intervals. You will find that at first there is no starch and that the intensity of the blue-black color of the test will increase with time as more sugar is synthesized into starch. (10) At the end of twenty-four hours, take a leaf from a region close to the stem of each of the seedlings and test for starch with the iodine solution. (11) The results should be as follows in the test tubes: #1, starch; #2, no starch; #3, no starch; #4, no starch; #5, starch; #6, no starch; #7, starch; #8, no starch.

Conclusion: We can thus conclude: (1) green cells make starch without sugar only in light; (2) albino plants, regardless of light, cannot make starch, thus showing that the chlorophyll of the green plants is necessary; and (3) green or not, light or dark, plant cells can only make starch if sugar is available.

Demonstrations: Use of Auxins

On Shoots

(1) Germinate soaked oat seeds in moist sand kept in the dark. When the coleoptiles emerge and are about 1.5 cm tall, cut off 3 mm of the tips of five of them with a sharp razor edge and apply a bit of lanolin paste to the cut surfaces. (2) Take another five cut coleoptiles and apply lanolin which has been treated with a 0.1% indoleacetic acid (100 ml of indoleacetic acid in 2 ml of absolute alcohol). (3) The third set of five coleoptiles should not be cut or treated. They should grow normally. (4) Place all three sets (cut tips in lanolin, cut tips with lanolin plus indoleacetic acid, and the normal tips) near a source of light. (5) After twenty-four hours, measure the angle of curvature in all three sets of plants. (6) The five treated with lanolin and indoleacetic acid will show the most variance.

On Roots

(1) Use germinated oat seeds but cut off 22 mm of several roots and use the same method as for the shoots. Then set up the seedlings in Petri dishes for geotropism demonstrations. The students will see that the auxins initiate root formation and roots elongate, but in this case the auxin inhibits further growth of roots.

On Cuttings

Dissolve 1.3 tsp. of indolebutyric acid in 1/2 tsp. of grain alcohol. Then stir this into five quarts of water, making a 0.01% solution. Place some geranium or coleus cuttings into some of this solution. Roots will be seen very shortly.

Demonstration: Gibberellin and Plant Growth

Use forty bean seeds that have been presoaked for several hours. Plant twenty of the seeds in vermiculite and label *experiment,* and the other twenty in vermiculite marked *control.* In about ten days select ten plants in the *experiment* box that are about the same size, label them from one through ten and discard the other ten in the box. Apply a drop of gibberellin solution, prepared as explained in a previous demonstration, to the shoot apex. Measure each plant in millimeters in the experimental and control groups. To the control plants apply a drop of water.

The control plants will be smaller and their growth slower than the gibberellin-treated plants.

Demonstration: Plant Growth Inhibition

Presoak bean seeds and then plant some in vermiculite. About ten days after the plants have grown to the stage where the first two leaves have expanded, select three that are about the same size. Tag one (1) phosfon, another (2) gibberellin, and the third (3) control. Apply a ring of phosfon-lanolin paste around the stem of plant #1 about a half-inch below the first pair of leaves and place plant in bright sunlight. Now apply gibberellin-lanolin paste to #2 and plain lanolin paste to #3. Use different applicators for each plant. Place all plants in the sunlight and examine every few hours for the next three days.

You will find that plants treated with phosfon are slower and will have thicker stems and darker green leaves than the control or the gibberellin plants. The plants treated with the gibberellin will be the tallest of the three plants.

AUDIO-VISUAL MATERIALS

FILMS

Behavior in Animals and Plants (K)
How Green Plants Make and Use Food (K)
Plant Life at Work (JJ)
Plant Tropisms and Other Movements (K)

FILMSTRIP

How Hormones Regulate Plant Growth (EE)

SINGLE-CONCEPT FILM LOOPS

Phototropic Response in Coleoptiles (Q)
Phototropism (HH)
Regulation of Plant Development: Coleoptile Response in ZEA (Q)

Molecular Genetics

UNIT 17

Although man did not know the actual mechanics, he did realize over the centuries that the parents of one species produced offspring of the same species. A horse did not give birth to a cat, nor did a cat give birth to a deer. The offspring tended to be like their parents in appearance and general characteristics.

Look at your own family and see whether there is any one member of the family who is completely different from the rest of the family. How does all this similarity come about? Biochemical genetics or molecular genetics is a relatively new branch of study.

The continuity of life—in cells, tissues, organs, and organisms—hinges on the process of reproduction. Although growth and metabolism are involved, over a period of time the continuation of life really depends upon the ability of the molecules, organelles, cells, and organisms to replace themselves.

You recall that the molecules present in the living organism vary all the way from simple inorganic substances to complex nucleoproteins. It is a simple matter to replace a simple molecule, but the more complex molecules are reproduced in a more complex manner.

It is a familiar fact to you that crystals grow through the accumulation of material from their surroundings by means of accretion, which is an increase in the volume and number of molecules in the particle through the acquisition of material from the outside. Molecular accretion is considered to be the simplest form of reproduction, shared by both living and nonliving materials.

Since many kinds of molecules are needed for replacement, and storage functions are not available from the environment, the organisms must manufacture these. These molecules cannot be made by merely mixing together two types of molecules; a special kind of reproduction machinery must be present inside the cell.

Each cell is a tiny chemical factory in which several thousand different kinds of chemical changes are constantly taking place among the numerous sorts of molecules that move about in its fluid or that are attached to its solid structures. These chemical changes are guided and controlled by the existence of many thousands of different enzymes within the cell.

We have just said that nearly every reaction that occurs in a living organism requires special enzymes. These act uniquely on fairly common molecules—glucose, fatty acids, glycerol, and amino acids, for example. Each species possesses its own peculiar set of enzymes. Feathered animals, in order to have feathers, have different enzyme systems from animals with fur. Whatever the genetic message may be, such as a "command" to develop as a deer, it seems likely that the message concerns the development of distinctive enzyme systems. We say that a certain characteristic, such as red hair, "runs in families"; that is so because enzymes for the production of red hair run in these families. Any theory concerning the chemical basis of heredity must explain how a species acquires and reproduces its special set of enzymes.

The catalytic action of enzymes brings about the formation of complex molecular units from simpler molecules. This is enzymatic synthesis and since the same enzyme molecule can be used again and again, it provides a means for reproduction of many units of the same substance.

Enzymes are specific in their activities, so the cellular machinery for enzymatic synthesis has a different enzyme for each of its synthetic reations. The specificity of enzymes is largely a matter of molecular shape. Molecules that can be joined together by an enzyme fit against the enzyme molecules as the pieces of a jigsaw puzzle fit together. Some molecules are so large and varied that they must have a master template system for their enzymatic synthesis. An enzyme molecule alone does not provide sufficient template surface to assure correct assembly of the molecule.

As was said before, enzymes are built up of some twenty different amino acids and since every different pattern of amino acid forms a molecule with its own set of properties, there are an enormous number of patterns possible. Of course, you realize that these large and varied molecules are the nucleic acids and proteins. Such large molecules form only when enzymes work in connection with the master template system.

Every cell is able to choose from among this vast number of possible patterns and select those that are characteristic of itself. Therefore it ends with a complement of specific enzymes that guide its own chemical changes and its properties and behavior. A fertilized ovum is "instructed" in the proper manner to develop and will choose a particular set of enzyme patterns out of all those possible.

The differences in this enzyme-guided behavior of the cells show in differences in body structure and thus in the make-up of different species. Even within a species different individuals will have slightly different characteristics, and this is accounted for by the slight distinction among their sets of genes. The only two humans that will be exactly alike will be identical twins who have originated as one fertilized ovum.

If the sperm and eggs of the parents contain the essentials to produce a unique enzyme system, then the germ cells of the children must possess these essentials too, in order that they, in turn, may pass them on to the next generation.

About 1900, Sir Archibald E. Garrod studied what he called *inborn errors of metabolism.* He believed that certain conditions, such as *alcaptonuria* or black urine of some children, result when there is an absence of certain genes. The person who (1) lacks a particular gene also (2) lacks a specific enzyme whose production the gene controls, and (3) is unable to carry out the metabolic reaction which the enzyme regulates. His concept is called the "one gene-one enzyme" hypothesis. However, it went unnoticed for almost fifty years until more investigations had been carried out and it was found that his hypothesis was correct.

As you are aware, each organism has chromosomes, important in the process of reproduction. Each chromosome can be considered as being composed of small reactions called genes. One thinks of genes being strung along the length of the chromosomes like beads. Each gene is considered to be responsible for the formation of a chain of amino acids in a fixed pattern, which is guided by the details of the gene's own structure. The structure is really the details of the gene's own structure, which can be translated into an enzyme's structure known as the genetic code.

The division of one parent cell involves one duplication of each gene, and this duplication is called replication. It would seem as if chromosomes are the substances in the cell capable of replication. If the genes furnished by the two parents for the zygote replicate in the development of germ cells in the offspring, then there is a basis for understanding how the units of heredity can be transmitted faithfully from one generation to the next.

The genes carry the genetic or heredity instruction. The message must be characteristic of the species; it must be faithfully and automatically passed from generation to generation; but at the same time it must be capable of evolutionary change when it occurs. When a particular enzyme or group of enzymes is formed imperfectly or not at all, there is the probability that there will be some visible abnormality of the body, such as color blindness, and the genes are referred to as "genes for color blindness."

It is also possible that a gene may exist in several varieties, such as a gene producing a slightly different enzyme; thus there might be a slight change in body characteristics. There is a gene for blue eyes and a gene for brown eyes; only one or the other, but not both, will be found in a specific place on a specific chromosome.

The two chromosomes of a particular pair govern identical sets of characteristics. Both have a place for the genes governing eye color. It might be possible that those on each chromosome of the pair may be identical, but both may be for blue eyes or for brown eyes, and in that case the individual is homozygous for that characteristic and is called a homozygote. It also could be that each chromosome of the pair may carry different varieties; the individual is then heterozygous for that trait and is a heterozygote. It is possible to be homozygous for one trait and heterozygous for another.

When an individual is heterozygous for a particular trait it can be that he shows the effect of only one of the gene varieties, and the effect will be from the dominant gene rather than the recessive. You recall that a gene having the necessary information to produce a given enzyme or other substance is referred to as a dominant gene. This gene will have a positive effect on the chemical output of the cell machinery, while a recessive gene is one which lacks the information to produce a given enzyme or other substance and therefore will impart a negative effect. If a cell has both a dominant and a recessive gene, the dominant gene will determine the resulting characteristic.

The human contains forty-six chromosomes in his somatic cells. However, through meiosis the sex cells have twenty-three, and when you stop to think that the fertilized ovum has twenty-three pairs of chromosomes, twenty-three from the sperm and twenty-three from the ovum, you will realize that a tremendous number of combinations can occur in the developing embryo. It is said that nearly ten million different combinations of chromosomes are possible in the sex cells of a single individual. Also, it is not possible to predict which chromosome combination in the sperm cell will end up in combination with which in the egg cell, so that by this reasoning a child can be produced with any of 100 trillion possible chromosome combinations.

This might explain the endless variety among living things, even within a particular species. Many of the variations result from the failure of parent cells to pass along certain genes.

DNA AND RNA

It has long been known that cell nuclei are high in polymeric material known as deoxyribonucleic acid, DNA. All evidence points to the fact that DNA constitutes the actual chemical of the genes. DNA is a member of a family of polymers called nucleic acids, and their monomer units are called nucleotides. The distinctness of any one nucleic acid (e.g., any one gene) lies in the order in which the nucleotide units are assembled and which ones are selected.

A single molecule can undergo self-duplication or replication. Within a living cell the nucleic acid, DNA, acts in this manner. The basis for self-duplication of cell parts, cells, and organisms is the ability of the DNA to produce exact copies of the original molecule. Another nucleic acid also develops from this DNA molecule template; this is ribonucleic acid (RNA). The RNA molecule is the basis for the protein-forming template of the ribosome.

In replicating, the DNA molecule must abide by the laws of conservation of matter and energy. All the starting materials which the molecule takes from the environment must be accounted for in the products and by-products of the process. The process must have a supply of nucleotides, the appropriate enzymes, and a source of energy to do the work building. All other biological or mechanical systems require the same things, but DNA is unique in that it provides its own creative pattern.

Prior to the 1950s, chemists had learned that DNA was a long molecule made up of subunits of phosphate, deoxyribose, and nitrogen-containing bases, adenine, guanine, cytosine, and thymine. (See Figure 17-1.) But in 1953, Watson and Crick, by means of X-ray diffraction data, proposed a model of DNA that showed it to have (1) a double-helix form, with the bases paired like rungs of a ladder; (2) the bases joined together by weak hydrogen bonds; and (3) the base adenine always paired with thymine, and guanine always paired with cytosine.

The most important part of cell division is the replication of DNA in the cell nucleus. According to Watson and Crick, this replication consists of a temporary uncoiling of the two helices of a DNA double helix, which is accomplished in a medium that contains fresh nucleotides. From this so-called mononucleotide pool, each uncoiled strand of DNA draws to itself those nucleotides that will fit with it via hydrogen bonds. Since those that fit are the same as those that are present at the same locations in the other, the now unzipped DNA strands of the original form a doubled helix. Each strand builds and twines about it a complementary strand, and the two new daughter helices again form a double helix which is identical with the parent in every respect.

When, in the presence of the necessary enzymes, these nucleotide subunits are added to each single helix to form the missing half of the double-helical structure of the DNA molecule, the arrangement of the four nitrogen bases provides the template pattern. (Figure 17-2.) The replication also requires auxiliary services from the cell. One of these services is an addition-elimination reaction whereby sugar and phosphate groups of adjacent nucleotides unite into the helical backbone of the DNA molecule. Of course, an appropriate enzyme is essential for this synthesis.

When a sperm cell unites with an egg cell at conception, the new cell, which is the zygote, has one new nucleus in which the DNA molecules from both parents intermingle and become the composite genes for the developing embryo. These genes (DNA

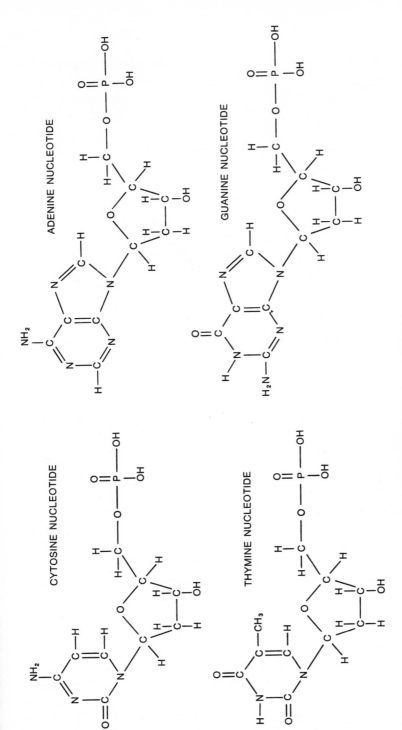

Figure 17-1 Basic nucleotides

231

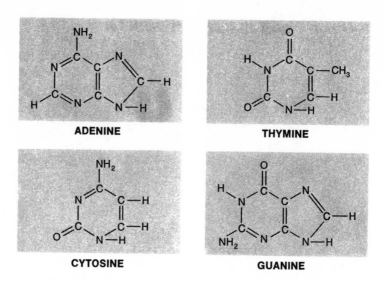

Figure 17-2 The four nitrogen-containing bases found in DNA

molecules) then direct the synthesis of both germ cells and somatic cells, so that the somatic cells develop into the member of the species of the parents and germ cells develop to be in readiness to transmit DNA molecules to the following generation.

The DNA molecule is coded for a specific genetic message, since the gene is a hereditary unit. All DNA have the same phosphate deoxyribose backbone, so that the difference between individual DNA molecules (genes) depends upon the order of amines that makes up the code of the genetic message a given gene bears. (Figure 17-3.)

In addition to serving as its own template, DNA also serves as a template for RNA. Just as the sequence of nucleotides in a single strand of DNA determines the sequence of a complementary strand, so also does this sequence determine the sequence of nucleotides in RNA. A different enzyme is involved in the addition-elimination reaction that joins the nucleotides of RNA. This very important enzyme is ribonuclease. RNA molecules help to control the synthesis of specific polypeptides, which will serve as enzymes.

Ribonucleic acid (RNA) is found throughout the cell, and differs from DNA in that it contains ribose instead of deoxyribose

AMINO ACID	TRIPLETS		
Alanine	GCU GCC GCA	Isoleucine	AUU AUC
Arginine	CGU CGC CGA	Leucine	UUA UUG CUU
Asparagine	AAU AAC	Lysine	AAA AAG
Aspartic acid	GAU ACG	Methionine	AUG
Cysteine	UGU UGC	Phenylalanine	UUU UUC
Glutamic acid	GAA GAG	Proline	CCU CCC CCA
Glutamine	CAA CAG	Serine	UCU UCG
Glycine	GUG GGC GGA	Threonine	ACU ACC ACG
Histidine	CAU CAC	Tryptophan	UGG
		Tyrosine	UAU
		Valine	GUU

Figure 17-3 Triplet messengers of the genetic code

and uracil instead of thymine. In the nucleus, RNA is found concentrated in two distinct regions: closely associated with the chromosomes, and in the nucleolus. In the cytoplasm, RNA is most highly concentrated in the ribosomes, the tiny bodies that dot the endoplasmic reticulum.

RNA is believed to have a helical structure quite similar to that of DNA.

The DNA unzips at the hydrogen bonds and, in presence of certain enzymes which catalyze the reaction and the ATP which provides the energy, the RNA nucleotides pair with the DNA bases, guanine to cytosine and uracil to adenine, and form single strands of RNA. After the RNA is formed on the DNA template, it peels off and leaves the exposed DNA nucleotides free to rejoin. It is obvious that the RNA formed by such a method is a complementary copy of the DNA upon which it was assembled. The complementary strand of RNA then moves out to the cytoplasm and, because it carries information from the DNA molecule in its structure, it is called *messenger RNA* or *mRNA*.

The miracle of how the cell consistently manages to assemble the amino acids in the proper sequence can easily be solved if you recall that enzymes and other proteins are made up of one or more polypeptide chains and that these chains are grouped together in a specific sequence. The sequence of amino acids is of critical importance, and if one amino acid is out of place in a polypeptide chain a different protein may result.

The sequence of nucleotides in the DNA molecule is a code that is somehow related to the proper sequencing of amino acids in polypeptides, and to another type of RNA called *transfer RNA*. A small section of a single strand of DNA may have the following sequences of bases: cytosine, adenine, cytosine, thymine, guanine, cytosine, adenine, adenine, guanine (C A C T G C A A G). Then the message in messenger RNA will have the following complementary sequence: G U U A C G U U C.

A three-nucleotide sequence within the DNA—for example, AAA—is copied by mRNA (UUU). Out in the cytoplasm this sequence attracts the mRNA, along with its amino acid phenyl-alanine, that has the exposed base sequence of AAA. In this manner all amino acids are placed in a sequence specified by the DNA sequence of nucleotides. A single letter code would allow for only four amino acids; a two-letter code would allow for only sixteen amino acids. Thus a three-letter code, allowing for sixty-four combinations would be more than adequate to code for twenty amino acids.

The main structural framework of a cell is the endoplasmic reticulum from which extend the granules called ribosomes, made of RNA and protein. About seventy-five percent of all the RNA of a cell may be found in these granules. Ribosomes are the sites of protein synthesis, but RNA does not itself appear to direct this work. Messenger RNA does this.

Messenger RNA molecules vary in length and may account for about five to ten percent of the total RNA in a cell. There is evidence that mRNA molecules can dissociate from ribosomes and that ribosomes can associate with different mRNA molecules of different molecular weight. There is also evidence that when protein synthesis takes place, it is much more difficult for mRNA to dissociate from ribosomes. There is evidence that a ribosome moves along a mRNA chain while the synthesis of a polypeptide, directed by mRNA, takes place. This synthesis requires that amino acids be brought to the site in the order in which they are to appear in the final polypeptide.

A characteristic is passed along from parent cells to the offspring by means of the gene. Occasionally a single gene through the DNA to RNA to protein synthesis pathway may result in the production of insufficient enzyme to produce the normal characteristic, and there will be a hereditary deficiency which will result in some sort of defect.

Some variations in the patterns of inheritance are not easily explained if we consider them in the light of fertilization, meiotic cell division, and the mechanisms of gene action, but can be explained if we assume that the DNA itself is altered. This alteration may be in the form of changes in the code of nucleotide sequence or of additions, deletions, and modifications of entire chromosomes or major parts of chromosomes.

Once there have been changes in DNA, they will be transmitted from generation to generation and thus will bring about a variation in the species. A change in the DNA molecule is a means by which biochemical conditions and structural characteristics are introduced into a species, and has a greater impact than a change which arises through a recombination of existing characteristics.

Biological evidence has long supported the conclusion that the least complex entity capable of self-duplication is the living cell. The DNA molecule is not self-sufficient; biological specificity is only partially responsible for this. DNA from one cell cannot be introduced into any kind of cell in a haphazard fashion with the idea that from that point on the cell must do what DNA tells it to. Carefully selected host cells can have their genetic apparatus taken over by foreign DNA, but such selection is necessary. Otherwise, the cell cannot and does not exhibit the expected changes. There must be present, for genetic specificity, not just the DNA, but also the appropriate enzymes and the nucleotides.

There is a relationship between a sequence of nucleotides on a gene (DNA) molecule and the sequence of amino acids in a protein. Since a protein synthesized under this genetic control would be the apoenzyme portion of an enzyme, it gives rise to the one gene-one enzyme idea.

There have been found in animal organisms several gene-to-enzyme-linked reactions that illustrate how important it must be that the transcription and translation of the genetic code be done, and done accurately. Phenylketonuria, PKU—excess phenylpyruvic acid circulating in the bloodstream—causes mental retardation. If discovered in infants and treated, it may prevent the onset of mental retardation. Two other hereditary defects are albinism and sickle-cell anemia.

The protein synthesis model has two advantages: (1) the gene, which always before had a vague definition, can now be defined precisely; and (2) mutations resulting in misplaced or misaligned amino acids can be accounted for. For example, 287 amino acids are found in one of the chains making up the hemoglobin molecule. Sickle-cell anemia can be caused by the substitution of a single amino acid in the 287 amino acids in each of the two chains.

A spontaneous development or change is the result of a mutation, which is an error in the duplication of the nucleotide code. With this alteration, the nucleotide code no longer spells out the characteristic normal for the species. The offspring may thus differ greatly from its parents. Among the mutagenic agents are extreme temperatures, various chemical compounds, electromagnetic radiations such as ultraviolet rays, X rays, gamma rays, and high-energy particles such as neutrons, beta rays, and cosmic rays.

It is only the mutations in germ cells which give rise to gametes, which may be transmitted to future generations. Most mutations are harmful in that they reduce the rate of survival. Any random change in an otherwise satisfactorily functioning machine is likely to be harmful. However, under certain conditions, a mutation may be of distinct advantage to a species.

All mutations are brought about by a chemical change, but the precise nature of the change is not established. A change in the chromosomal pattern also brings about a variation in inheritance. Some of the changes are aberrations in the formation of single chromosomes. Crossing over is one such change which may occur during the first division of meiosis, when paired chromosomes (each consisting of two strands of DNA joined at the centromere) come to lie in adjacent positions along the equator of the spindle. Since the DNA strands are joined only at the centromere, the remainder of these lengthy molecules may become intertwined. If a strand from one chromosome is crossed over a strand of the other chromosome of the pair, the sugar-phosphate bonds of each double helix may break at the point of crossing. The pieces of the two DNA strands can unite to form a new gene combination. With movement of the chromosomes to the poles of the cell, these altered chromosomes are separated, destined to become incorporated in different gametes. Thus, genes that were formerly linked in a distinct genetic pattern may have become separated from the genes with which they are normally associated. They are now linked in a new package, with the possibility of new genotypic and phenotypic combinations.

Much of the work in genetics has been concerned with the determination of specific inherited traits and the mode of inheritance. Such studies often call for records of many generations of related individuals. This, of course, has proven to be a problem in the study of human inheritance. Few families have detailed knowledge of their more remote ancestors. Now, however, new approaches to genetic study have become possible. With knowledge of the structure and coding system of the DNA molecule, precise information at a fundamental level can be obtained.

(Use models to show "unzipping" as well as many films, filmstrips, and single-concept film loops.)

Laboratory Exercise: Human Heredity

Aim: To determine the frequency of certain human traits in a population.

Although human heredity is studied primarily by means of pedigrees or family records, there are several traits that can be studied immediately to help you understand human heredity. They are tongue rolling, PTC tasting, earlobes, and eye color.

Materials Needed: PTC (Phenylthiocarbamide) papers; your immediate family, friends, and classmates.

Procedure:

(A) Inheritance of Eye Color

Make a survey of fifty persons to determine their eye color. Dark eye color consists of brown or black pigments while light eye color consists of gray, blue, or green pigments. Record your findings as to color and sex. In your family, which is inherited? Would you say that the sex of an individual is related to eye color?

(B) Inheritance of Earlobes

Make a survey of fifty persons to determine whether their earlobes are free or attached to the jaw. Record your findings as to type of earlobe and sex. In your family, is the trait inherited? Would you say that the sex of an individual is related to earlobe attachments?

(C) Inheritance of Tongue Rolling

Make a survey of fifty persons to determine their ability or inability to roll the tongue. Record your findings as to sex. In your family, which is inherited? Would you say that sex of an individual is related to the ability to roll the tongue?

(D) Inheritance of the Ability to Taste Phenylthiocarbamide

Make a survey of fifty persons to determine the ability of an individual to taste PTC as a bitter sensation against the inability to taste PTC. It is said that about seventy percent of the population can taste the PTC. Record your findings as to the ability to taste PTC and sex correlation.

Conclusions: Since you have used the members of your family for the various tabulations, would you say that two different children of the same parents could have traits different from each other? Why?

Laboratory Exercise: Action of Genes in Peas

Aim: To show that differences which seem to be just gross may well be microscopic and chemical as well. Thus the biochemical and enzymatic differences of the peas are hereditary and of importance in many ways in the matter of breeding.

Materials Needed: Thirty smooth peas; thirty wrinkled peas; two small bottles; medicine droppers; mortar and pestle; razor blade; Petri dishes; balance; compound microscope; slides; cover glasses; paper towels; distilled water; nutrient agar; iodine solution for starch tests; blender; funnel; filter paper; cheesecloth; one gram glucose-1-phosphate.

Procedure:

(A) Gross Differences

(1) Weigh the thirty smooth peas and then the thirty wrinkled peas. Record their weight. Now place the peas into two

bottles, one bottle labeled for the smooth peas and one bottle labeled for the wrinkled peas. Fill the bottles almost full of water and put aside until the next day. (2) The following day, remove the peas from the water and dry them as well as possible with paper towels. Now weigh them and record the weight after soaking for twenty-four hours. Determine the weight increase in both sets of peas and calculate the percentage of increases in weight. (3) What can you now say about the appearance of both kinds of peas and the ability of each to soak up water?

(B) Microscopic Differences

(1) Take a microscope slide and mark one-half of the slide with an "S," for smooth, and the other with a "W," for wrinkled. Place a drop of water on either side of the slide. (2) Cut through a soaked smooth pea with a razor blade and scrape a little of the cut surface into the drop of water on the half of the slide marked "S." Do the same with the soaked wrinkled pea. Mix thoroughly with water. (3) Use the low power of the microscope to examine the scrapings then use the high power. Describe and draw what you see. There should be a definite difference in the starch grains, with the smooth seed having smooth, undivided starch grains as contrasted with the divided, dimpled starch grains of the wrinkled peas.

(C) Chemical Differences

(1) Make a 2% agar solution which should contain .5% glucose-1-phosphate. One grain is sufficient for twenty Petri dishes. Boil the agar and the glucose until the foam looks coarse. You do not have to sterilize the mixture but pour into Petri dishes to form a layer about 5 mm deep. Allow to gel. Cover the dishes and refrigerate until needed. (2) Weigh out twelve grams of dry smooth peas and twelve grams of dry wrinkled peas. (3) Either grind each group to a fine powder with a mortar and pestle or place into a blender for a few seconds to get a fine powder. (4) Now place the powder into a bottle, add 12 ml of distilled water and mix thoroughly. Filter through several layers of cheesecloth. Label each bottle either smooth or wrinkled filtrate. (5) Take the Petri dishes, and divide the plates into two halves by drawing a line with a wax pencil on the outside of the bottom dish. Mark one side "S" and the other "W." (6) With a medicine dropper, place four separate drops of the extract from the smooth seeds on the side marked "S" and then place four separate drops of the extracts from the wrinkled seeds on the side marked "W."

If you have an incubator, place the dishes in it for thirty minutes. Otherwise, allow them to stand for thirty minutes on the table. This is so that the extract can act on the glucose. (7) At the end of that time, add a drop of iodine to two of the drops on each side of the dish. You should find that a blue ring has formed at the drops of the smooth seed extract and a very dark blue ring at the wrinkled seed extract. Now wipe off the extract and iodine with filter paper. (8) Allow the dishes to stand for another thirty minutes. Test the remaining two drops on each side of the dish with the iodine solution. There should be very little or no starch found in the smooth peas, while there still are traces of starch in the extract of the wrinkled peas.

Since we are assuming that both the smooth and wrinkled peas have been dried under similar conditions, the reason for the differences in the gross features, the type of starch grains found, and the enzymes controlling the synthesis of carbohydrates must come from the genetic differences of the peas.

AUDIO-VISUAL MATERIALS

FILMS

  Biochemical Genetics (Parts I and II) (B), (H)
  Chromosome Chemistry and Genetic Activity (B), (H)
  DNA Structure and Replication (B), (H)
  Gene Action (B), (J)
  Gene Structure and Gene Action (B), (H)
  Genetics: Function of DNA and RNA (K)
  Mitosis and DNA (K)
  Mystery of Life (H)

FILMSTRIP

  DNA, A Key to All Life (X)

SINGLE-CONCEPT FILM LOOPS

  DNA Structure: Backbone and Bases (Q)
  DNA Structure: The Meselson and Stahl Experiment (Q)
  DNA Transformation Experiment (Q)
  Extracting DNA (AA)
  Fruit Fly Chromatography (A)

Giant Chromosome to Drosophila (A)
Identifying the Genetic Material (Q)
Identifying Genetic Material: Phage Experiment (Q)
Molecular Basis of Heredity (Q)
Properties of Genetic Material (Q)

TRANSPARENCIES

Biochemical Basis of Genetics (set of five) (O)
Cell Nucleus and DNA (F)
Comparison of DNA and RNA (F)
Deoxyribonucleic Acid (F)
DNA Set (A)
Interesting Facts About DNA (F)
Replication of RNA (F)
Ribonucleic Acid (F)
RNA and Protein Synthesis (F)
Structure of DNA (F)

CHARTS

DNA (F), (P)
RNA (F), (P)

MODELS

Bacteriophage (F), (CC)
DNA (CC)
DNA Molecule Construction Kit (S)
RNA Protein Synthesis (F), (O), (CC)
Student DNA Model Kit (F), (S)

KITS

Blood Typing Kits (F), (S), (CC), (DD), (EE), (FF)
Rh Typing Kits (F), (S), (CC), (DD), (EE), (FF)

PAMPHLETS

DNA: The Secret of Life (X)
Facts of Life (II)
Molecular Genetics (JJ)

APPENDIX

Sources of Audio-Visual Materials

(A) Wards Natural Science Establishment, P.O. Box 1712, Rochester, New York 14603

(B) Lifelong Learning, University Extension, University of California, Berkeley, California 94720

(C) Filmstrip of the Month, 355 Lexington Avenue, New York, New York 10017

(D) Sargent-Welch Scientific Co., 7300 Lindner Street, Skokie, Illinois 60076

(E) Bailey Film Associates, 11559 Santa Monica Blvd, Los Angeles, California 90025

(F) CCM: General Biological Inc., 8200 South Hoyne Avenue, Chicago, Illinois 60620

(G) Denoyer Geppert, 5235 Ravenswood Avenue, Chicago, Illinois 60640

(H) McGraw-Hill Films, 330 W. 42nd Street, New York, New York 10036

(I) Association Films, 600 Madison Avenue, New York, New York 10022

(J) Encyclopedia Britannica Education Corp. 425 North Michigan Avenue, Chicago, Illinois 60611

(K) Coronet Films, 65 E. Water Street, Chicago, Illinois 60601

(L) Rand McNally Company, Box 7600, Chicago, Illinois 60680

(M) National Teaching Aids Inc., 120 Fulton Avenue, Garden City, New York 11040

(N) Colab Laboratories, Inc., 3 Science Road, Glenwood, Illinois 60425

(O) La Pine Scientific Company, 6001 South Knox Avenue, Chicago, Illinois 60629

(P)   Hubbard Scientific Company, 2855 Sherman Road, North-brook, Illinois 60062

(Q)   Ealing Films, 2225 Massachusetts Avenue, Cambridge, Mass-achusetts 02140

(R)   Stanley Bowmar Company, Inc., Valhalla, New York 10595

(S)   Science Kit, Inc., Tonawanda, New York 14150

(T)   Carolina Biological Supply Company, Burlington, North Carolina 27215

(U)   Macalaster Scientific Company, Nashua, New Hampshire 03060

(V)   Society for Visual Education, Inc., 1345 Diversey Parkway, Chicago, Illinois 60614

(W)   Doubleday Multi-Media, 277 Park Avenue, New York, New York 10017

(X)   Life Education Program, Box 834, Radio City P.O., New York, New York 10019

(Y)   Association Films, 512 Burlington Avenue, La Grange, Illinois 60525

(Z)   Modern Learning Aids, 1212 Avenue of Americas, New York, New York 10036

(AA) Thorne Films, Inc., 1229 University Avenue, Boulder, Colorado 80302

(BB) Harper & Row Publishers, 44 E. 33rd Street, New York, New York 10016

(CC) Stansi Scientific, 1231 N. Honore Street, Chicago, Illinois 60622

(DD) Lab-Pak Scientific Corp., P.O. Box 297, Baldwin Place, New York 10505

(EE) Lab-Aids, Inc., 160 Rome Street, Farmingdale, New York 11735

(FF) Wilkens-Anderson, 4525 West Division Street, Chicago, Illinois 60651

(GG) Indiana University, Audio-Visual Center, Bloomfield, Indiana

(HH) Popular Science Publishing Company, 355 Lexington Avenue, New York, New York 10017

(II) Eli Lilly and Company, Indianapolis 6, Indiana

(JJ)   Moody Institute of Science, 11428 Santa Monica Blvd., Los Angeles, California 90024

(KK)   Tweedy Transparencies, 208 Hollywood Avenue, East Orange, New Jersey 07018

Index

A

Acetic acid as organic compound, 34
Acidosis in blood, 188
Acids, bases and salts, 57-68
    acids, 57-58
        Arrhenius definition, 58
        Brönsted-Lowry definition, 58
        properties, 58
    audio-visual materials, 67-68
    bases, 59
        definition, 59
        properties, 59
    hydrogen ion concentration, 60-61
        demonstration, 64-67
        importance of pH, 62-63 (*see also* "Hydrogen ion concentration")
        indicators, 61-62
        pH term, 60-61
    neutralization, 59-60
        demonstration, 64
    salts, 60
Adenine as nitrogenous base of DNA, 133-135
Adenosine triphosphate, 144-148
Aerobic respiration, 144-146
    of glucose, 144-146
Aerosols, 50
Agglutinogens, 187
Albinism, 236
Alcohol as simplest organic compound, 33
Alcoholic fermentation, 148
Aldehydes as types of organic compound, 33
Alpha carbon of amino acids, 113
Alpha-tocopherol, 179-180
Amines as organic compounds, 34-35

Amino acid, electrophoresis of, demonstration of, 124-125
Amino acids, proteins and, 112-118
Amylase, demonstration of, 219-220
    optimum pH for functioning of, 167-168
Amylopectin, 102
Amylopsin, 162
Amylose, 102
Anaerobic respiration, 146
Anaerobic respiration of yeast, laboratory exercise of, 150-152
Analysis, 39
Analysis of protein as laboratory exercise, 123-124
Animal digestion, 154-171 (*see also* "Digestion, animal")
Animals, cellular respiration in, 142-144
Apoenzyme, 86-87
Arginine, 113
Arrhenius' definition of acid, 58
Ascorbic acid, 178-179
    test for, 188-189
Atomic mass and atomic number, difference between, 28-29
Atomic structure of matter, 27-29
    and element, difference between, 29
    mass and number, difference between, 28-29
ATP, 144-148
Auxins, 216-217
    demonstration of use of, 222-223

B

Barfoed's test for sugar, 104
Bases, 59